DIGITAL FILTERING
A Computer Laboratory Textbook

THE GEORGIA TECH DIGITAL SIGNAL PROCESSING LABORATORY SERIES

Editors:

Thomas P. Barnwell III

Monson Hayes

Russell M. Mersereau

Mark J. T. Smith

Texts in this series include:

Introduction to Digital Signal Processing: A Computer Laboratory Textbook
by Mark J. T. Smith and Russell M. Mersereau

Digital Filtering: A Computer Laboratory Textbook
by Russell M. Mersereau and Mark J. T. Smith

Spectral Analysis: A Computer Laboratory Textbook
by Monson Hayes

Speech Coding: A Computer Laboratory Textbook
by Thomas P. Barnwell III, Kambiz Nayebi, and Craig H. Richardson

Georgia Tech
DIGITAL SIGNAL PROCESSING
LABORATORY SERIES

DIGITAL FILTERING
A Computer Laboratory Textbook

RUSSELL M. MERSEREAU

MARK J. T. SMITH

Georgia Institute of Technology

WILEY

JOHN WILEY & SONS, INC.
New York • Chichester • Brisbane • Toronto • Singapore

Acquisitions Editor	Steven Elliot
Marketing Manager	Debra Riegert
Senior Production Supervisor	Savoula Amanatidis
Cover Designer	Bonnie Cabot
Illustration Coordinator	Sigmund Malinowski
Manufacturing Manager	Andrea Price

This book was typeset in Times Roman by the authors and printed and bound by Malloy Lithographing, Inc. The cover was printed by Phoenix Color Corp.

Library of Congress Cataloging in Publication Data:

Mersereau, Russell M.
   Digital filtering : a computer laboratory textbook / Russell M.
Mersereau, Mark J.T. Smith.
      p.   cm.
   System requirements for computer disk: IBM-compatible PC (80286
microprocessor or better recommended); 640K RAM; MS-DOS; hard disk;
CGA, EGA, or VGA display; floating point math co-processor and
standard ASCII text editor recommended.
   Includes bibliographical references (Pref.).
   Includes index.
   ISBN 0–471–51694–5
   Electric filters, Digital–Design and construction–Data
processing. 2. Signal processing–Digital techniques–Data
processing. 3. Digital filters (Mathematics) I. Smith, Mark J. T.
II. Title.
TK7872.F5M467   1993
621.3815'324'078–dc20   93–17580
                                                   CIP

Foreword

After spending decades in the research laboratory, digital signal processing (DSP) is now emerging to make a significant impact on many areas of technology. As a result, DSP is becoming a basic subject in the electrical engineering curriculum. Although numerous textbooks and reference books are available to present the theory and applications of DSP, few of these books provide much in the way of "hands-on" experience that can help a student translate equations and algorithms into insight.

Experience during the past fifteen years at the Georgia Institute of Technology in using computers with both basic and advanced courses in DSP has shown that the personal computer can be an extremely effective learning aid when it is combined with well-designed exercises and effective software support. The Georgia Tech Digital Signal Processing Laboratory Series builds on this teaching experience to provide a set of computer laboratory books that can be used either to supplement traditional classroom/textbook presentations of the subject or as a self-study aid.

The value of computer-based laboratory experience is clear. However, just what this experience should be is somewhat dependent on the computer resources available and on the computer skills of the students. The following three approaches have proved to be effective:

1. Provide the student with a program or set of programs that perform specific DSP functions. In this situation, exercises are necessarily limited to running the programs on test data and observing the results.

2. Provide the student with a set of exercises that can be carried out by using a set of macros or low-level functions that can be strung together in some sort of convenient software environment. This approach has the virtue of flexibility and is much less restrictive.

3. Provide the student with test data and suggestions for projects to be carried out with

whatever programming resources are available. Clearly, this is the least restrictive approach, but is the most demanding of the student's programming/computer skills.

The first approach is likely to be frustratingly limited for students who are learning fundamental concepts, but it is very appropriate when the goal is to demonstrate complex algorithms that would require a great deal of time if students were to implement them on their own. For example, digital speech processing systems often combine many basic DSP functions and often have many parameters whose effects can only be illustrated and studied by using an elaborate program. Another example is filter design, where students can learn the properties of different approximation methods by simply applying those methods to the same set of specifications. At the opposite extreme is the third approach, which is obviously most suited for advanced courses or independent study where appropriate computer programming skills can be required. The second approach is perhaps the best compromise for developing insight into the fundamental algorithms and concepts of DSP. The book *Digital Filtering: A Computer Laboratory Textbook*, is based on primarily this approach and is the second book in the Georgia Tech Digital Signal Processing Laboratory Series. It addresses the set of topics related to filter design, implementation, and analysis and is a follow-on to the first book *Introduction to Digital Signal Processing: A Computer Laboratory Textbook*.

Digital Filtering includes more than 170 exercises that can be carried out under DOS using carefully designed software provided with the book. This software has a wide range of basic operations, a large set of filter design functions, and a structure that allows these functions to be strung together to perform more complex functions. At Georgia Tech, this computer laboratory mode of operation has been underway for several years. Student response to both the software and the exercises has been extremely favorable. Students appreciate the ease with which they can begin to actually do something with what they are learning in the classroom.

There is no doubt that DSP education is moving toward the greater use of computers. Indeed, few subjects in the electrical engineering curriculum are so well suited to the use of computers in instruction. The Georgia Tech Digital Signal Processing Laboratory Series, whose authors have many years experience in teaching and research in the DSP field, is a valuable contribution to this emerging trend in electrical engineering education.

Ronald W. Schafer
John O. McCarty Institute Professor
Georgia Institute of Technology

Preface

Filter design is an important topic in the area of discrete-time processing. There are many digital signal processing textbooks that present a good discussion of the theory, provide a variety of illustrative examples, and include a wide selection of homework problems related to the major topics. The purpose of this book, however, is somewhat different. It is to provide hands-on exposure to digital filter design in a computer environment. It can be used to complement a digital signal processing (DSP) text, as the text for an introductory *laboratory* course in digital signal processing, or as a self-paced introduction to DSP basics.

The book includes a library of DSP computer functions that run on personal computers using the DOS operating system. The philosophy underlying this text is to provide the DSP newcomer with the experience of working with complex design formulas and design algorithms without having to write and debug large programs. Computer-based exercises have been a very important component in the digital signal processing course offerings at Georgia Tech and strongly contribute to an enriched understanding of the material.

This book is the second in a two-part sequence that focuses on the fundamental concepts of digital signal processing. The first book, *Introduction to Digital Signal Processing: A Computer Laboratory Textbook*, covers linear systems, the discrete Fourier transform, sampling, the z-transform, the DFT and FFT, and certain other topics. This text is a continuation and is devoted to digital filters, digital filter design, and filter implementation. It assumes that the reader is familiar with most of the topics covered in the first text. Each chapter begins with a brief summary of the fundamentals on its topic. These discussions are followed by a set of illustrative exercises that provide a mix of theoretical, experimental, and design problems. Many of the problems are straight-forward, and their solutions can be verified easily and quickly by using the computer. The exercises also include a number of more difficult problems to challenge the learner. Certain exercises can be selectively omitted without a loss of understanding, according to the reader's level of familiarity with and interest in a particular topic.

The text is organized so that proceeding through the initial chapters and exercises in order provides a smooth introduction to the software that is used throughout the text. The presentation is most effective when chapters are selected in order but chapters may be selectively omitted without loss of continuity. It is suggested, however, that Chapter 1 be reviewed first. It reviews the fundamentals of FIR and IIR filters, establishes the notation, and provides an introduction to the software. Chapter 2 treats FIR filter design. It includes discussions and exercises on window design, the Remez exchange algorithm, and the Parks–McClellan algorithm. Chapter 3 introduces classical analog filter design methods, analog frequency translation, a variety of analog to digital filter conversion methods, and digital frequency transformations. Chapter 4 discusses allpass filters, their properties, and some fundamentals of multirate filtering. Chapter 5 is devoted to filter structures and implementation issues. Coefficient quantization, roundoff errors, and limit cycles are treated in some detail. The final chapter of the book is a collection of projects. These use the concepts and techniques discussed throughout the text to solve specific problems. A topical reference chart is included in this preface that shows how the chapters in this book and its predecessor would best complement those in several commonly used DSP textbooks.

The software used in this book is included on the enclosed disk. There are three basic executable programs: **f.exe**, which contains a diverse set of elementary signal processing and filter design functions; **g.exe**, which contains all of the display graphics functions; and **helpf.exe**, the on-line help for the **f** and **g** functions. The software will generally run on any personal computer that supports the DOS. It has been primarily tested on IBM PS/2 and HP Vectra PC systems. It is strongly recommended that the computer supporting the DSP software have a floating-point coprocessor. The programs will also run more efficiently if all of the software and data files reside on a hard disk. The graphics functions should operate properly on most EGA-, CGA-, and VGA-equipped machines.

In the development of this laboratory text we were fortunate to have received valuable feedback from colleagues and students at Georgia Tech in the United States and Georgia Tech Lorraine in France. We gratefully acknowledge Prof. James McClellan for his contributions to Chapter 6 and Dr. Steven L. Eddins for his development of the early core set of computer programs that evolved into the present software. The software has undergone much revision and modification during the course of its development. We are indebted to Mr. Faouzi Kossentini and Mr. Wilson Chung who over the last few years have revised, maintained and expanded the software as the book developed. We would like to gratefully acknowledge the following reviewers for their suggestions on changes and improvements in the manuscript: Matt Yeldin, University of California, Berkeley; Ernie Baxa, Clemson University; Enrico Del Re, Universita Delgi Studi Di Firenze; Frederick J. Harris, San Diego State University; and Janet Slocum, Tufts University. Finally, we would like to thank our wives Martha Mersereau and Cynthia Y. Smith and our children Adam and David Mersereau and Stephen, Kevin, and Jennifer Smith for their steadfast love and support over the years.

This book and its companion were written to be laboratory texts and not to be the primary text in a lecture course. Several popular primary texts are listed below. These can be relied upon for more complete discussions, examples, and derivations of the key results that we have only summarized. The tables indicate which chapters in the primary texts provide overlapping coverage with the chapters in these two laboratory texts.

[1] L. B. Jackson, *Digital Filters and Signal Processing*, (2nd), Kluwer Academic Publishers: Boston, 1989.

[2] R. Kuc, *Introduction to Digital Signal Processing*, McGraw-Hill: New York, 1988.

[3] L. C. Ludeman, *Fundamentals of Digital Signal Processing*, Harper and Row: New York, 1986.

[4] A. V. Oppenheim and R. W. Schafer, *Discrete-Time Signal Processing*, Prentice-Hall: Englewood Cliffs, NJ, 1989.

[5] A. V. Oppenheim and R. W. Schafer, *Digital Signal Processing*, Prentice-Hall: Englewood Cliffs, NJ, 1975.

[6] J. G. Proakis and D. G. Manolakis, *Introduction to Digital Signal Processing*, Macmillan: New York, 1988.

[7] R. A. Roberts and C. T. Mullis, *Digital Signal Processing*, Addison-Wesley: Reading, MA, 1987.

[8] R. D. Strum and D. E. Kirk, *First Principles of Discrete Systems and Digital Signal Processing*, Addison-Wesley: Reading, MA, 1988

Intro. DSP (Smith & Mers.)	Ch.1	Ch.2	Ch.3	Ch.4	Ch.5	Ch.6
1) Jackson	——	Ch.2	Ch.4	Ch.6	Ch.3	Ch.7
2) Kuc	——	Ch.2	Ch.3	Ch.3	Ch.5	Ch.4
3) Ludeman	——	Ch.1	Ch.1	Ch.1	Ch.2	Ch.6
4) Opp. & Sch., 1989	——	Ch.2	Ch.2,5	Ch.3	Ch.4	Ch.8,9
5) Opp. & Sch., 1975	——	Ch.1	Ch.1	Ch.1	Ch.2,4	Ch.3,6
6) Proakis & Manolakis	——	Ch.2	Ch.4	Ch.1	Ch.3	Ch.9
7) Roberts & Mullis	——	Ch.2	Ch.4	Ch.4	Ch.3	Ch.4,5
8) Strum & Kirk	——	Ch.3	Ch.4	Ch.2	Ch.5,6	Ch.7,8

Dig. Filters (Mers. & Smith)	Ch. 1	Ch. 2	Ch. 3	Ch. 4	Ch. 5
1) Jackson	——	Ch. 9	Ch. 8	Ch. 13	Ch. 5, 11
2) Kuc	——	Ch. 9	Ch. 8	——	Ch. 6, 10
3) Ludeman	——	Ch. 3	Ch. 3,4	——	Ch. 6, 10
4) Opp. & Schafer, 1989	——	Ch. 7	Ch. 7	Ch. 10	Ch. 6
5) Opp. & Schafer, 1975	——	Ch. 5	Ch. 5	——	Ch. 4,8
6) Proakis & Manolakis	——	Ch. 8	Ch. 8	——	Ch. 10, 7
7) Roberts & Mullis	——	Ch. 6	Ch. 6	——	Ch. 9, 10
8) Strum & Kirk	——	Ch. 9	Ch. 10	——	Ch. 11

Contents

Introduction 1

Digital filters can be used in a wide variety of applications, including separating signals from noise, compensating for linear distortions, separating signal components that have been added together, and modeling many classes of signals. This book presents a series of computer exercises in digital filtering to acquaint you with a number of filter design techniques and methods of filter implementation. It comes with custom DSP software that will allow you to design filters, study design algorithms, and study many types of digital filters without having to write programs. Although many of the fundamentals are discussed in this book, it is not intended to be a basic text in digital signal processing or in digital filtering. Instead, its goal is to supplement such a text by providing a hands-on "computer laboratory" experience.

This chapter has two purposes; first, it reviews some basic concepts related to digital filters while establishing the notation used throughout the book. It also provides a smooth introduction to the use of the software. Chapters 2 and 3 present a series of tutorial exercises on various aspects of FIR and IIR filter design, and Chapter 4 looks at some more advanced topics concerned with allpass and multirate filters. Chapter 5 explores alternative implementations for digital filters and carefully examines quantization and overflow effects that can occur in fixed-point hardware realizations. The text concludes with a series of design projects in Chapter 6 that are more lengthy in nature and require a higher level of thought and creativity.

1.1 GETTING STARTED

In this text, signal processing operations are presented in a hands-on personal computer environment that can create and display signals with only a few commands. To get started you will need the following:

1

- An IBM-compatible personal computer. A computer with at least an 80286 or 80386 microprocessor is preferred, although not necessary. It is also recommended that the computer contain a floating-point co-processor. This will increase the speed of the software dramatically.

- A hard disk. Since files will be created routinely when you do the exercises, it is suggested that a few megabytes of disk be available in your working directory.

- The MS-DOS operating system or its equivalent.

- Either CGA, EGA, or VGA display capability.

- The computer should contain a minimum of 640 kbytes of random access memory. Since the DSP software uses a sizable part of this memory, avoid running other memory resident programs during your work session with the software. The presence of these programs in memory reduces the memory available to the DSP software and may cause errors.

A standard ASCII text editor is also useful. It will allow you to write macros and to edit DSP files. In Chapter 6, which is the projects chapter, and in two optional problems in Chapter 5, you are asked to write computer programs. In such cases, you will need a compiler for the programming language in which you wish to work.

A *print-screen* program that allows you to make hard copies of the graphics displayed on the screen might also be useful in some cases, but is not necessary. The DSP software does not provide the capability to print graphics outputs. You will generally be asked to draw sketches of signals that are displayed on the screen.

To begin, create a DSP directory on your hard disk and copy all of the files on the enclosed diskette into that directory. Go into the DSP directory and type **install**. The programs and files provided on the disk are stored in a compressed format on the diskette. Typing **install** uncompresses the software. It is recommended that you copy these programs and files to a backup diskette as a safeguard in case they are accidentally deleted or overwritten.

You may do all of your work in this DSP directory. However, you may find it more convenient to work in another directory, thereby keeping your working files separate from the DSP software. Such a setup can be created by modifying your search path in DOS. Your DOS manual contains detailed information about customizing the operating environment for your computer.

There are two main programs in the software that contain DSP functions: **f.exe**, which contains filter design and signal processing functions; and **g.exe**, which contains graphics and display functions. The functions in **f.exe** can be used to do simple operations such as adding, subtracting, or multiplying signals as well as to perform more complex operations such as filter design, multirate filtering, and quantization simulations. Each function in this set can be invoked by simply typing **f** followed by the function name (e.g., **f add, f subtract, f multiply**).

The functions in **g.exe** allow you display signals in the time domain, z-plane, and frequency domain. These graphics functions are invoked by typing **g** followed by the function name. A list of the basic functions for this book is provided in Table 1.1.

Table 1.1.   List of DSP Software Functions.

f functions			
f abessel	f abutter	f acheby1	f acheby2
f adaptfir	f add	f aelliptic	f atransform
f bartlett	f bilinear	f blackman	f cartesian
f cas	f cexp	f convert	f convolve
f diff	f direct1	f direct2	f divide
f dnsample	f dtransform	f eformulas	f extract
f fdesign	f fft	f filter	f gain
f hamming	f hanning	f hilbert	f histogram
f ideallp	f ifft	f imagpart	f impinv
f kaiser	f kalpha	f lccde	f log
f lshift	f mag	f matchedz	f maxflat
f median	f multiply	f nlinear	f obutter
f ocheby1	f ocheby2	f oelliptic	f par
f phase	f pksmcc	f polar	f qcyclic
f quantize	f rank	f realpart	f reverse
f revert	f rgen	f rootmult	f rooter
f siggen	f snr	f subtract	f summer
f truncate	f upsample	f zeropad	

g functions			
g afilspec	g afreqres	g apolezero	g dfilspec
g dtft	g look	g look2	g polezero
g slook2	g sview2	g view	g view2

Working with the software is very simple and does not require knowing much in order to get started, but there are several things that we should point out before you begin. First, signals (or sequences) are stored in files. Their content may be examined at any time by simply printing them on the screen, i.e., by entering **type** followed by the filename. As an example, try typing **f001**, which is a file provided for you on disk. After pressing the enter key, the file content will be displayed. Observe that the first five lines of the file provide information about the signal while the numbers that follow are the sequence values. This file format is convenient for modification because the file information is self-explanatory. Using a text editor, you can change coefficient values if desired as well as add or delete coefficients. In the case of the latter, the filter length and numerator order parameters would have to be changed appropriately. As another example, consider the file **impulse**, which contains the unit sample, $\delta[n]$. It is used as an input signal in many of the exercises. Type this file to the screen and observe that it contains only one sample. The fact that the starting point is zero means that the unit impulse occurs at $n = 0$.

Second, whenever you are in doubt about what a particular function does or how to use it, simply type **helpf** for an on-line description of the functions in **f.exe** and **g.exe**. Try typing **helpf** now. It will display a list of all the functions available. To obtain detailed information about a particular function listed, type **helpf** followed by the function name.

Try typing **helpf add** for an example of the on-line help feature.

Third, the graphics and display functions are all contained in the **g.exe** program. When a plot is being displayed, pressing the "esc" key will return you to the main menu, or in those cases where there is no main menu, it will return you to the operating system. Pressing the "q" key exits the program. As an example, try typing **g view f001**, which will display the sequence stored in f001. There are two graphics functions that are slightly different: **g polezero** and **g apolezero**. These are completely menu driven and prompt you for all information.

Fourth, IIR filters (which are discussed later in this chapter and in Chapter 3) have the form $H(z) = B(z)/A(z)$. These files are stored with the numerator and denominator coefficients listed separately. To illustrate this, type **f003** to the screen and observe that the numerator and denominator coefficients are easily identifiable. More is said about the file structure for IIR filters in Exercise 1.4.4.

Fifth, to invoke any of the DSP functions, just type the function name. You will then be asked for any required arguments, such as the name of the input file, the name of the output file, and any appropriate function parameters. Alternatively, these can be included on the command line as shown below:

$$\textbf{f function}\quad\textbf{arg1}\quad\textbf{arg2}\;\dots$$

The ordering of the arguments will vary from function to function, but normally the input file(s) are listed first, followed by the output file(s), and then any floating point or integer parameters that are required. The program will ask for any arguments that you omit. For example, consider the function **f add**, which has two inputs and one output. Assume that $x[n]$ is the signal stored in the file **f001**. The operation

$$y[n] = x[n] + x[n]$$

can be implemented by typing

$$\textbf{f add}\quad\textbf{f001}\quad\textbf{f001}\quad\textbf{yn}$$

where **yn** is the output file containing $y[n]$. Try this and then display **yn** using **g view**. Remember that the *"esc"* or *"q"* keys will allow you to exit the function.

Finally, sequences can be complex valued. Complex numbers are stored as ordered pairs containing the real and imaginary parts. For example, the complex number $2.5+j5.3$ is represented by the pair 2.5 5.3. Notice that the real and imaginary parts are separated by a space. Manipulating complex sequences is similar to manipulating real ones. For example, the operation

$$y[n] = (2 + j2)x[n]$$

can be realized using the **f gain** function. Type

$$\textbf{f gain}\quad\textbf{f001}\quad\textbf{yn}$$

where again we assume $x[n]$ is the signal contained in **f001**. You will be prompted for the value of the gain. Specify **2   2** corresponding to the complex number $2 + j2$. Alternatively, the value of the gain can be specified on the command line by typing

$$\textbf{f gain}\quad\textbf{vn}\quad\textbf{yn}\quad\textbf{2}\quad\textbf{2}$$

Try displaying **yn** using **g view**. Notice that now a menu appears requesting options regarding how you wish this complex sequence to be displayed.

1.2 WRITING AND USING MACROS

Functions can be listed sequentially in a file and can be executed in order by calling that file name. Such files are called *batch files* in DOS and have a **.bat** extension. In other computing environments, this sequential call of programs is viewed as macrolevel programming and the files containing them are called *macros*. Following this viewpoint, we will refer to these files as macros in this book.

Several macros are included with the software for your convenience. They are **df1.bat**, **scale.bat**, and **qpoly.bat**. Documentation describing their function and usage is included in the first few lines of the files. It can be viewed by typing the macro to the screen.

When a sequence of operations is repeated several times in the exercises, you may wish to write these commands in a macro so that the sequence of commands can be performed using a single command. To illustrate how to write a macro, consider creating one that performs a scaling operation on an IIR filter. Specifically, the macro is to receive as input an IIR filter of the form $H(z) = B(z)/A(z)$ and change its amplitude by an arbitrary amount, α. This is equivalent to multiplying the numerator polynomial, $B(z)$, by α while keeping the denominator polynomial, $A(z)$, the same. To do this, the macro first splits the IIR filter into two sequence files, one containing $A(z)$ and the other containing $B(z)$, using the **f convert** function. It then uses the **f gain** function to multiply the numerator file by the constant and reassembles the modified numerator and original denominator into one file again using **f revert**. This macro has been included in the software as an example. Type the macro **scale.bat** to the screen at this time. Notice that the first few lines describe its operation and usage. The list of commands that follow implement the procedure just described. Two points regarding input/output are important to mention at this time. A macro can receive input or return output from the command line by using the %1, %2, %3, ... specifications. DOS substitutes the first file listed on the command line for every occurrence of %1 in the macro. Similarly the second is substituted for %2, the third for %3, and so on, thus allowing direct input and output from the command line. As an illustration of how this works, examine the macro, **scale.bat**, by typing it to the screen. You will observe that the first few lines contain a description of the macro. These lines are ignored at the time of execution because each begins with a colon ":" symbol. Text on a line that begins with a colon is ignored by DOS.

To further illustrate the operation of the macro, consider the IIR filter stored in the file **f003**. Use the macro to scale this filter by a gain factor of 2. Do this by typing

$$\textbf{scale} \qquad \textbf{f003} \qquad \textbf{out1}$$

where **out1** is the IIR file that will contain your output. The macro executes each function in sequence. When it comes to the **f gain** function, you will be prompted for the gain value.

The second point to mention is that parameters can also be specified on the command line of the macro. For example, by using another command line argument, %3, in the macro the gain parameter can be specified directly. Copy the **scale.bat** macro into another file called **scale1.bat** and modify the new macro to accept the gain on the command line.

This can be done with a text editor. The only change involved is to add the %3 to the line containing the gain function. Thus the modified line would be

$$\textbf{f gain} \quad \textbf{\_temp} \quad \textbf{\_temp} \quad \textbf{\%3}$$

Try repeating this example by typing

$$\textbf{scale1} \quad \textbf{f003} \quad \textbf{out2} \quad \textbf{2}$$

Type **out1** and **out2** to the screen and verify that they are the same.

Note that several intermediate files are generated in the macro. In the case of **scale.bat**, they are **_temp** and **_temp1**. To avoid accumulating unnecessary files in your directory, the macro deletes these files after they are used. The command **del _temp?** deletes all files of the form **_temp?** where "?" is any alphanumeric character recognized by DOS. To avoid the potential of having a macro overwrite an important file in your directory, it is suggested that you adopt the convention of never beginning a file name with an underbar "_". Reserve the underbar for temporary files that you use in your macros or for holding intermediate results. Such a convention will help avoid accidental overwriting of files.

As a final comment, we mention that the **scale.bat** function is just included as an example and not as a macro intended for use. The **f gain** function when applied to IIR filters actually scales the filter in an identical manner and it is much faster. Thus **f gain** should be used for all scaling.

Much more can be said about macros and how to write them. Additional information on this subject is available in Section 1.3 of the companion text, *Introduction to Digital Signal Processing: A Computer Laboratory Textbook* and in the DOS manual under batch files.

The next few sections of this chapter contain a very basic discussion of filters and have two main purposes. The first is to review some fundamental concepts that should be familiar. The second is to provide a smooth introduction to the software.

1.3 FIR AND IIR FILTERS

Filters are usually divided into two classes: FIR (**F**inite length **I**mpulse **R**esponse) filters and IIR (**I**nfinite length **I**mpulse **R**esponse) filters. IIR filters contain feedback, whereas FIR filters do not. As a result the nth output sample, $y[n]$, of an FIR filter is a function of only a finite number of samples of the input sequence, whereas the nth sample of an IIR filter may depend upon an infinite number of input samples. FIR filters are most often implemented directly using a relation similar to the convolution sum, although they can also be implemented using discrete Fourier transforms (DFTs). IIR filters are implemented recursively using difference equations.

FIR filters have a number of features that make them attractive in applications.

- FIR filters have good numerical properties when implemented using fixed-point arithmetic.

- FIR filters can be designed to have exactly linear phase characteristics.

- FIR filters are extremely well suited to many multirate filtering implementations and can demonstrate significant improvements in computational efficiency in these applications.
- FIR filters are always stable.
- High-order FIR filters can also be implemented efficiently using the DFT.

IIR filters also have a number of desirable attributes.

- IIR filters of low order can be designed that possess excellent frequency response magnitude characteristics.
- IIR filters can be designed directly from classical analog filters.
- IIR filters are often better suited for applications involving signal modeling and system simulation than FIR filters.

In the remainder of this text, these issues will be expanded and illustrated through the computer exercises.

1.4 FREQUENCY SELECTIVE FILTERS

Many kinds of discrete systems can be viewed as digital filters, but linear, time-invariant, frequency selective filters are the ones most frequently considered in the context of filter design. These are filters that pass only selected regions of the frequency spectrum. Several types of frequency selective filters are shown in Figure 1.1. These vary in the form of their magnitude responses. Notice that in some regions the frequency response is unity while in other regions it is zero. These filters are said to be *ideal filters* because they do a perfect job of either passing or eliminating a region of the spectrum. Unfortunately, ideal filters cannot be implemented using a finite amount of computation, in general.

Real digital filters differ from ideal filters in several respects. The passbands of real filters are not perfectly flat. Practical stopbands can attenuate, but cannot reject bands of frequencies completely. The transition between passband and stopband behavior cannot be instantaneous, but must take place over a band of frequencies, called the *transition band*. The more closely a real filter resembles an ideal filter, the better the quality of the filter. Filter quality can be measured in terms of several parameters, many of which are illustrated in Figure 1.2. The degree to which the filter deviates from its ideal passband and stopband is called the *maximum passband and stopband deviation*, denoted δ_p and δ_s, respectively. Because real digital filters often contain ripples in the passband and stopband, δ_p and δ_s are often called the *passband* and *stopband ripple*. Another specification for the maximum stopband deviation is the attenuation, ATT, defined as

$$\text{ATT} = -20 \log_{10} \delta_s.$$

It is measured in decibels (dB). An equally important characteristic of a filter is the width of its transition band, $\Delta\omega$, and its center or nominal cutoff frequency, ω_c. For an ideal filter, the transition width is zero. However, for realizable filters, this width lies somewhere between 0 and π radians (rad). The edges of the transition band are ω_p and ω_s and are called the *passband* and *stopband cutoff frequencies*. Thus, ω_p and ω_s

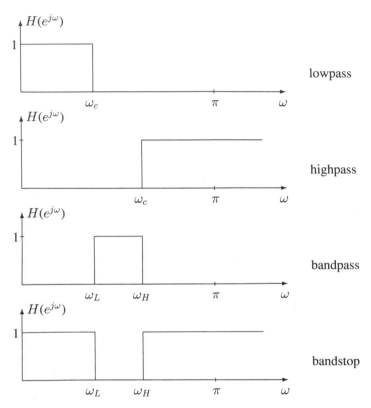

Figure 1.1. Some common frequency selective filters.

provide the same information as $\Delta\omega$ and ω_c. These cutoff and deviation parameters are illustrated in Figure 1.2 for a lowpass filter.

The last group of parameters relates to the size of the filter, i.e., the number of zeros and poles that it contains. The *order*, N, is equal to the number of poles in the filter's system function. The complexity of an FIR filter is related to its length, which is one greater than the number of zeros that it contains. Thus there are three groups of design parameters that characterize a real filter: (1) the passband/stopband deviation parameters, which include δ_p, δ_s, and ATT; (2) the transition width parameters, ω_p, ω_s, or $\Delta\omega$ and ω_c; and (3) the filter order or, in the case of FIR filters, the filter length.

These are the parameters that are commonly specified in the filter design problem. They apply to lowpass, highpass, bandpass, multiband, and bandstop filters. For bandpass filters, bandstop filters, and filters with many passbands (multiband filters), transition widths and maximum deviation parameters must be specified for each transition region and passband or stopband region. It is worth mentioning that there are more sophisticated design methods that allow for the specification of phase or group delay characteristics. However, these are beyond the scope of this text.

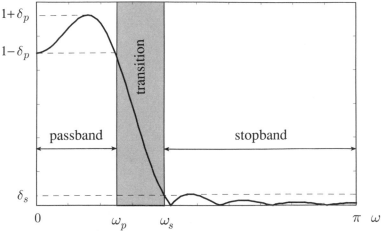

Figure 1.2. Magnitude response specifications for a digital lowpass filter.

Although the major thrust of our treatment is on digital filters, we will also discuss the design of classical analog filters. This topic is relevant because our IIR design approach is based on designing analog filters first and then converting (mapping) these into digital filters. The conventional design parameters or design specifications for analog filters are slightly different. This is discussed in greater detail in Chapter 3. The primary difference is that the maximum value of the filter's magnitude response is restricted to be less than or equal to one. The passband gain thus has a maximum value of 1 and a minimum value of $1 - \Delta_p$ as shown in Figure 1.3. The remaining parameters are similar. The maximum passband and stopband deviations are denoted Δ_p and Δ_s, respectively, and the corresponding attenuation is ATT $= -20 \log_{10} \Delta_s$ (dB). The transition width and center cutoff frequency are defined the same as for the digital case and are denoted $\Delta\Omega$ and Ω_c, respectively. The passband and stopband cutoff frequencies are Ω_p and Ω_s as shown in Figure 1.3.

1.5   THE SYSTEM FUNCTION AND FREQUENCY RESPONSE

The *system function* of an IIR filter is the z-transform of its impulse response. It is a rational function of the complex variable z^{-1}.

$$
\begin{aligned}
H(z) &= \frac{z^L \sum_{k=0}^{M} b_k z^{-k}}{\sum_{k=0}^{N} a_k z^{-k}} \\[2em]
&= \frac{b_0 z^L \prod_{k=1}^{M} (1 - z_k z^{-1})}{a_0 \prod_{k=1}^{N} (1 - p_k z^{-1})}.
\end{aligned}
$$

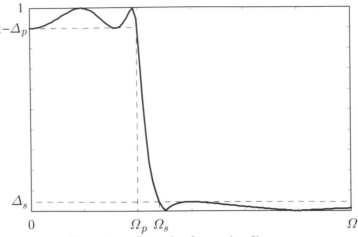

Figure 1.3.   Example of an analog filter.

The roots of the numerator polynomial, $\{z_k\}$, are called the *zeros* of the system and the roots of the denominator polynomial, $\{p_k\}$ are called the *poles*. These roots may be real or complex. The number of poles, N, (apart from those at the origin or at infinity), is called the *order* of the filter. The special case $N = 0$ results in an FIR filter. The complexity of an FIR filter is normally measured by its *length*, $M + 1$. On occasion, the term *order* is used in association with an FIR filter. The order of an FIR filter is the number of zeros or equivalently the order of its z-transform polynomial.

The *frequency response* corresponds to the special case of the system function when the complex variable, z, is restricted to the unit circle.

$$H(e^{j\omega}) = H(z)\big|_{z=e^{j\omega}}$$

Because the frequency response is complex, it is often plotted using its real and imaginary parts

$$H(e^{j\omega}) \stackrel{\triangle}{=} \underbrace{H_R(e^{j\omega})}_{real\ part} + j\ \underbrace{H_I(e^{j\omega})}_{imag.\ part}$$

or, more commonly, using its magnitude and phase

$$H(e^{j\omega}) \stackrel{\triangle}{=} \underbrace{|H(e^{j\omega})|}_{magnitude}\, e^{j\,\overbrace{\arg[H(e^{j\omega})]}^{phase}}$$

or its log magnitude and group delay, $\tau(\omega)$.

$$\text{logmag}[H(e^{j\omega})] \quad \stackrel{\triangle}{=} \quad 20\log_{10}|H(e^{j\omega})|$$

$$\tau(\omega) \quad \stackrel{\triangle}{=} \quad -\frac{d}{d\omega}\arg[H(e^{j\omega})].$$

These functions are shown in Figure 1.4 for the simple first-order system $H(z) = 1/(1 - 0.9e^{-j\omega})$.

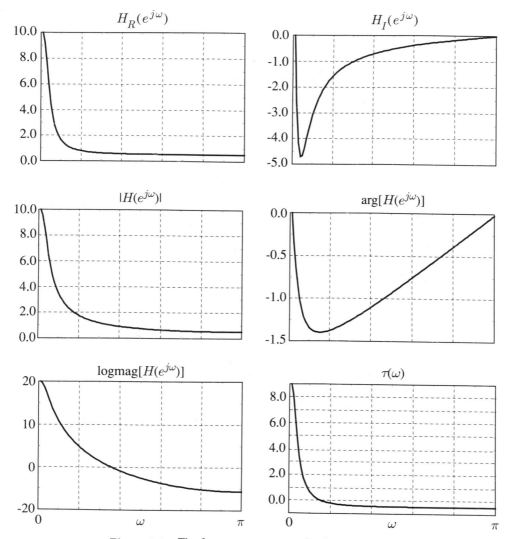

Figure 1.4.   The frequency response of a first-order system.

EXERCISE 1.5.1. **Introduction to Using the Software**

You can invoke a DSP function simply by typing its name. To get started, we shall design an FIR lowpass filter $h[n]$ using the function **f fdesign**. This function designs filters based on a technique called *windowing*—a method we shall discuss later in Chapter 2. Type **f fdesign** and press the "enter" key. The program will ask you for an output file name. This is the file that will contain the impulse response samples of $h[n]$. Call your output file, **hn**. It is strongly suggested that you choose meaningful file names, as this will help you keep track of your filters. Next, the function will ask for the filter length. For this exercise, specify 10 as the length. The function will then ask whether you want a lowpass or a highpass filter. You should select a lowpass filter for this exercise. The next prompt is for the center cutoff frequency which, in this exercise, should be $\pi/2 = 1.57$. The final selection involves a choice of windows. Choose the rectangular window. This is equivalent to an ideal lowpass filter with its impulse response truncated so that its total length is 10.

To examine the filter, type **g view hn** and press the "enter" key. This will graphically display the impulse response of your filter. Pressing the "esc" key will allow you to exit the program. As an additional exercise in using the software, display **hn** using the functions **g view2** and **g sview2**. These functions allow you to display two filters or signals at the same time: one below the other in the case of **g view2** and; one superimposed on the other in the case of **g sview2**. Try this by typing

$$\textbf{g} \qquad \textbf{view2} \qquad \textbf{hn} \qquad \textbf{hn}$$

and

$$\textbf{g} \qquad \textbf{sview2} \qquad \textbf{hn} \qquad \textbf{hn}$$

You can also examine the frequency response of the filter by typing **g dtft hn** and pressing the "enter" key. A menu will appear asking for a selection of different plots. Starting with the first item on the menu, examine each plot. The "esc" key should be used to return to the main menu after you have finished viewing a plot. Notice that the function **g dtft** displays the frequency plots on a frequency axis that extends from $-\pi$ to $+\pi$ rad and thus the origin $\omega = 0$ appears in the middle of the screen.

You may also examine the filter by typing **g polezero**. This is a menu-driven function that allows you to view the pole/zero plot, the magnitude response, the filter coefficients, and list the pole and zero locations. It also allows you to move, delete, or add poles and zeros and write your results to an output file. First, type **g polezero**. Enter "1" to read a file and specify **hn** as that file. Observe that a message appears at the top of the screen indicating that some of the zeros are outside the field of view. When this occurs, option 2 can usually be used to adjust the size of the plot so that all of the roots are displayed. Adjust the circle size at this time. Now display the numerical values of the zeros using menu option 3.

Return to the main menu. Next, try changing the location of a zero by specifying option 4. Enter 2 in the next submenu, indicating that you want to change the location of a zero. The program provides the flexibility to move roots individually or in complex conjugate pairs as is evident by the menu selection. Try moving the zero located at $z = -1$. To do this, use option 5, which specifies a single zero. The number corresponding to the zero at $z = -1$ is 1 and should be entered next. Now use the directional arrows to move that zero. Press the *"d"* key to see the new magnitude display after each move. The increment by which you move the root can be changed by using the *"Page Up"* and *"Page Down"* keys. Increase the increment to 0.1 by pressing the *"Page Up"* key. You may wish to experiment a little bit moving roots around and exploring the many features of the **g polezero** function. To exit this part of the function, press the return key. This will return you to the submenu. □

EXERCISE 1.5.2. Log Magnitude Plots

Use **f fdesign** to design an FIR lowpass filter $h[n]$ of length 64 with a cutoff frequency of 1.0 rad using a *Hamming* window. Display and sketch its magnitude response using **g dtft**. The fine details of the behavior of the stopband response are difficult to see on this plot because the stopband ripples are so small. This is typical of many good filters that we will consider in later chapters. To get a more insightful picture of the ripple behavior in the stopband, we often use a log magnitude plot, which is a plot of the function $20 \log_{10} |H(e^{j\omega})|$. The log magnitude plot is one of the options listed on the **g dtft** menu. Press the *"esc"* key and examine the log magnitude plot of your filter and sketch it. This representation has the property that a unity magnitude passband has an attenuation of zero. More importantly, on a log magnitude plot the stopband amplitude range extends from zero to negative infinity. Thus the stopband ripples that are very small become clearly visible. The key to interpreting the log magnitude response quickly is to remember the correspondence between the log scale (in the log magnitude plot) and the linear scale (in the magnitude plot). The log magnitude scale is in units of decibels (dB), like the attenuation, so the attenuation can be easily determined from the log magnitude plot. What is the stopband attenuation of $h[n]$? What is the value of the maximum stopband deviation, δ_s, if the attenuation, ATT, is 20 dB? What is δ_s, if ATT equals 40, 60, and 80 dB? □

EXERCISE 1.5.3. Evaluating Filter Characteristics

When we want to compare two filters, we usually compare their orders, their transition widths, and their stopband and passband deviations. Many digital filters have ripples in their passband and stopband that make these parameters unambiguous. By simply examining the magnitude response, a unique set of parameters δ_p, δ_s, ω_p, and ω_s can be measured. Use **f fdesign** to design a length 32 lowpass filter file, **hn**, with a center cutoff frequency of 1.5 rad using the *rectangular window* option. Display and sketch the magnitude response of this filter using **g dtft**. By inspection, estimate δ_p, δ_s, ω_p, and ω_s. In general it is difficult to determine these parameters accurately by a simple inspection of the plot. The **g dfilspec** function is provided to make this task easier. Type

$$\textbf{g} \qquad \textbf{dfilspec} \qquad \textbf{hn}$$

and select a display of the magnitude response. Press the right directional arrow key *"→"*. A cursor and pop-up window will appear, which display the frequency ω at the cursor location, the magnitude $|H(e^{j\omega})|$ and $1 - |H(e^{j\omega})|$. The right- and left-arrow keys move the cursor position right and left in very small increments. The up- and down-arrow keys move the cursor right and left in large increments. Try moving the cursor across the plot using the directional arrow keys. Use this feature to obtain a more accurate estimate δ_p, δ_s, ω_p, and ω_s.   □

EXERCISE 1.5.4.   Creating Filter Files

The software contains many functions that will design a wide variety of filters and put them in files. However, on occasion, you may want to create your own filter and enter its coefficients in a file that can be read by other functions. The *create file* option in **f siggen** will do this. As an example, consider the IIR filter with a system function

$$H(z) = \frac{b_0 + b_1 z^{-1} + b_2 z^{-2} + b_3 z^{-3}}{1 + a_1 z^{-1} + a_2 z^{-2} + a_3 z^{-3}}, \tag{1.1}$$

where $b_0 = -0.09, b_1 = -0.113, b_2 = -0.113, b_3 = -0.09, a_1 = -1.37, a_2 = 1.07,$ and $a_3 = -0.305.$

(a) Use the function **f siggen** to create a file called **hn** with these coefficients. The function will prompt you for all of the necessary information. Note that the filter is IIR, the numerator and denominator orders are 3, and that the coefficients are real. Specify the starting point to be zero. This means that the filter impulse response begins at zero. Enter **hn** for the file name and enter the coefficients. Display the magnitude and log magnitude responses of this filter using **g dtft** and sketch these plots.

(b) The coefficients of IIR filters are stored as lists of numerator and denominator coefficients. The leading coefficient of the denominator is always assumed to be one and is therefore omitted in the representation. Print the coefficients on the screen by typing

type hn

Notice that the numerator and denominator coefficients are listed in order. In addition, the first five lines of the file (called the *header*) contain information about the degrees of the numerator and denominator polynomials, whether the coefficients are real or complex, and the starting point of the impulse response. The first parameter, which is called the *length,* is the total number of coefficients listed in the file.

Create a file **gn** for the FIR filter with system function $G(z)$, where

$$G(z) = 1 + 2z^{-1} + 2z^{-2} + z^{-3}$$

using the *create file* option of **f siggen**. Note that the filter is FIR, its length is 4, the coefficients are real, and that the starting point is 0. Display and sketch the magnitude and log magnitude responses using **g dtft**. As before, type the coefficients of this file to the screen. Observe that the coefficients are treated as a sequence and appear as a list immediately following the header.   □

EXERCISE 1.5.5. **Monotonic Filters and the 3-dB Point Convention**

When filters have magnitude responses with ripples in the passband and stopband, these ripples imply a logical partitioning of the spectrum into its passband, stopband, and transition band regions. However, for filters with a monotonic response, this is not true. As an example, consider the filter $H(z)$, where

$$H(z) = -0.125 - 0.575z^{-1} - 1.0z^{-2} - 0.75z^{-3} - 0.125z^{-4} + 0.125z^{-5} + 0.05z^{-6}.$$

Use the *create file* option in **f siggen** to create a file **hn** for this filter. Display its magnitude response using **g dfilspec** and sketch the plot. The monotonically decaying magnitude response does not provide natural definitions for $\delta_p, \delta_s, \omega_p$, and ω_s. Thus either the passband and stopband deviations or the passband and stopband cutoff frequencies must be specified initially. Then the other parameters can be determined.

In such cases, the 3-dB cutoff frequency is a popular parameter. It can be used as the edge of the passband for a filter response or it can be used as the nominal cutoff frequency. It is the frequency where the magnitude response drops to $\sqrt{2}/2$ of its nominal passband value. For filters with a unity gain passband, the gain at the 3-dB point is approximately 0.707. Use the directional arrow keys to move the cursor and display the peak magnitude of $H(z)$. Remember that the up- and down-arrow keys move the cursor in large increments while the right- and left-arrow keys move in small increments. Estimate the 3-dB cutoff frequency (in radians) for the filter $H(z)$.

A popular convention for defining the stopband deviation δ_s in the case of a filter with a monotonic magnitude response is to let δ_s equal one-tenth the value of the nominal passband gain. For the case of a unity gain filter, $\delta_s = 0.1$. Assuming these conventions for the passband and stopband deviations, use **g dfilspec** to determine the 3-dB cutoff frequency, and the stopband cutoff frequency for the filter $H(z)$. □

EXERCISE 1.5.6. **A Complex Filter**

The coefficients of most filters used in practice are real numbers. However, these coefficients can be complex when the need demands it.

(a) Use **f fdesign** to design a length 20 FIR lowpass filter, $H(z)$, with cutoff frequency of 1.5 rad using a Hanning window. Display the filter's magnitude response using **g dtft** and sketch the plot.

(b) Use **f cexp** to form the new complex filter

$$g[n] = h[n]e^{j\omega_0 n}$$

with $\omega_0 = 1$. Type the coefficients to the screen and notice that they are complex. Recall that complex numbers are represented in the software as ordered number pairs. Use **g dtft** to display the magnitude response of $g[n]$ and sketch the plot. The magnitude response is no longer symmetric in frequency about $\omega = 0$.

(c) Examine $H(z)$ and $G(z)$ using **g polezero**. Adjust the circle size to be small so that all zeros appear in the field of view. Carefully sketch the pole/zero plots for both filters. What is the relationship between the zeros of $G(e^{j\omega})$ and $H(e^{j\omega})$? □

EXERCISE 1.5.7. **Effect of Poles and Zeros on the Magnitude Response**

The position of the poles and zeros in the z-plane has a well-understood effect on the magnitude response. A zero near the unit circle causes a dip in the magnitude response at that frequency, and a pole near the unit circle causes a peak in the response. To investigate this effect, run the **g polezero** function and invoke its *change pole/zero* option. Specifically, add a pole at $z = 0.5$ using option 4. Then return to the main menu and add a zero at $z = -1$. Use the *move zero* and *move pole* options to move the pole and zero closer to and farther from the unit circle. Pressing the "d" key will automatically update the magnitude response plot. Pressing the "enter" key will return you to the submenu. Describe the behavior that you observe. What happens when the zero is placed on the unit circle? What happens when the pole is placed on the unit circle? What happens when the pole and zero are placed on top of each other? □

EXERCISE 1.5.8. **Estimating the Poles and Zeros of a Filter**

The rules of thumb discussed in the previous exercise can be used to estimate the locations of poles and zeros from the magnitude response plot of a filter. Consider the file **f001**, which contains the impulse response coefficients of a low-order FIR filter. Examine the magnitude response of this filter using **g dtft**.

(a) By looking at this plot, estimate the location of the zeros of the filter that lie on the unit circle.

(b) Now estimate the locations of the zeros that do not lie on the unit circle.

(c) Estimate the length of the filter. Remember that the length of an FIR filter is related to the number of zeros it has.

(d) Repeat this exercise for the FIR filter in file **f002**. It may be helpful to examine the log magnitude response of this filter as well.

(e) Now consider the low-order IIR filter in file **f003**. Estimate the number of poles in the filter and try to estimate their locations.

(f) Finally, examine the magnitude response of the low-order filter contained in the file **f004**. This filter has both poles and zeros. Estimate their locations.

Use **g polezero** to verify the accuracy of your estimates for each of the filters. □

EXERCISE 1.5.9. **Sketching the Magnitude Response**

In this exercise, we practice applying the rules for sketching the magnitude response from the pole/zero plot.[1]

Listed below are the poles and zeros for four filters, $H_0(z), \ldots, H_3(z)$.

---

[1]This procedure is discussed in detail in Chapter 5 of the companion book *Introduction to Digital Signal Processing: A Computer Laboratory Textbook*. It is also a common topic in basic DSP texts and is often called the geometric interpretation of the magnitude response.

Filter	Poles	Zeros
$H_0(z)$:	none	j
		$-j$
		1.0
$H_1(z)$:	$j0.9$	j
	$-j0.9$	$-j$
$H_2(z)$:	0.9	1.0
	$0.769 + j0.558$	1.0
	$0.769 - j0.558$	1.0
$H_3(z)$:	$0.141 + j0.893$	$0.715 + j0.700$
	$0.141 - j0.893$	$0.715 - j0.700$
	$-0.141 + j0.893$	$-0.715 + j0.700$
	$-0.141 - j0.893$	$-0.715 - j0.700$

Sketch the pole/zero plots for each of these filters and their corresponding magnitude response plots based on the locations of the poles and zeros. Classify each of them as lowpass, highpass, etc., and estimate their center cutoff frequencies. The computer is not required for this exercise. □

1.6 DIFFERENTIATORS AND HILBERT TRANSFORMERS

Two other ideal filters, which appear frequently in applications, are differentiators and Hilbert transformers. When the input to an ideal differentiator consists of samples of a bandlimited continuous time signal, the output will be samples of its derivative. Figure 1.5 displays the frequency response of an ideal differentiator. A differentiator is usually designed as a linear-phase FIR filter.

The real and imaginary parts of a complex analytic signal are Hilbert transforms of each other. An ideal Hilbert transformer, which will evaluate samples of the Hilbert transform of a bandlimited signal, is really a digital filter with the purely imaginary frequency response shown in Figure 1.6. These are useful components in a number of modulation systems, some of which are explored in the design problems in Chapter 6.

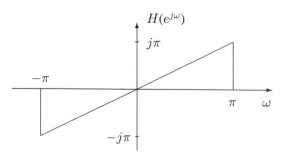

Figure 1.5. The frequency response of an ideal differentiator.

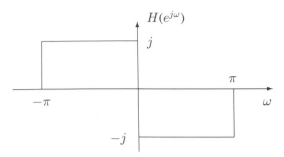

Figure 1.6.   The frequency response of an ideal Hilbert transformer.

EXERCISE 1.6.1.   A Differentiator

The function **f diff** produces N samples of the impulse response of an ideal differentiator. Use this function to design an 8-point differentiator, **hn**. This exercise considers the effect of this filter on a triangular sequence.

(a) Examine the frequency response of the differentiator using **g dtft** and sketch the magnitude and phase responses. Next examine and sketch the pole/zero plot using **g polezero**.

(b) Generate a 41-point triangular signal file **xn**, using the *triangular wave* option in **f siggen** by specifying the period to be 40, the number of periods to be one, the amplitude to be 1, and the starting point to be 0. Display the triangular sequence using **g view** and sketch the plot.

(c) Use **f convolve** to convolve the triangular wave sequence with the differentiator and put the output in the file **yn**. Display the original **xn** and the output **yn** together by typing

g view2   xn   yn

Sketch this plot and observe that **yn** resembles the derivative of **xn** appropriately delayed.   □

EXERCISE 1.6.2.   Hilbert Transformer

An ideal Hilbert transformer has the frequency response shown in Figure 1.6. A real Hilbert transformer only approximates this ideal. The function **f hilbert** designs real Hilbert transformers. Use it to design a 15-point Hilbert transformer **hn**. Shift **hn** by 7 samples to the left by typing

f lshift   hn   hn   −7

This will center the impulse response about zero. Display the real and imaginary parts of the frequency response using **g dtft** and note that the real part is approximately zero. Sketch the pole/zero plot and the magnitude response using **g polezero**. Notice that since the magnitude response is positive, we see symmetry about $w = 0$. Observe the ripple behavior of the filter as it attempts to follow the steplike shape of the ideal Hilbert transformer.   □

FIR Filter Design 2

Finite length Impulse Response (FIR) filters play an important role in the design of practical discrete-time systems. Unlike IIR filters, they can be designed to be causal and have exactly linear phase, a condition that is explored in the next section, and they can be implemented using the discrete Fourier transform. This chapter will look at several different algorithms for the design of FIR filters including the window method, frequency sampling, maximally flat FIR filter design, and the Parks–McClellan algorithm. Each of these design approaches has its advantages. The window method is particularly simple, the frequency sampling method ties the design parameters to values of the discrete Fourier transform, and the Parks–McClellan algorithm designs filters that are optimally *equiripple* in the minimax, or Chebyshev, sense. We will also look at some related FIR filtering techniques, including adaptive FIR filters and some nonrecursive nonlinear filtering techniques that have become widely used. The discussion of FIR filters continues beyond this chapter. FIR filters are discussed in the context of multirate systems in Chapter 4, and implementation issues relevant to these filters are addressed in Chapter 5. An alternate design method for FIR filters that are optimal in the mean-square error sense is introduced in Chapter 6.

2.1 LINEAR-PHASE CONDITION

A *linear-phase* filter is one whose frequency response can be written in the form

$$H(e^{j\omega}) = e^{-j\lambda\omega}\hat{H}(e^{j\omega}),$$

where $\hat{H}(e^{j\omega})$ is either real or purely imaginary. All linear-phase filters have impulse responses that display some kind of symmetry. If $h[n]$ is the impulse response of an

N-point causal, linear-phase FIR filter, it must satisfy the condition

$$h[n] = \pm h^*[N - 1 - n]. \tag{2.1}$$

The complex conjugate appears in (2.1) for completeness, but $h[n]$ is usually real. Thus, unless otherwise stated, we will assume $h[n]$ to be real. When (2.1) is true with the plus sign (which is the symmetric case), $\hat{H}(e^{j\omega})$ is real; when it is true with the minus sign (the antisymmetric case), $\hat{H}(e^{j\omega})$ is purely imaginary. The linear-phase parameter $\lambda = (N - 1)/2$ represents the point about which the impulse response is symmetric (or antisymmetric). When N is odd, the point of symmetry is an integer corresponding to one of the sample values; when it is even, the point of symmetry lies between two sample values.

EXERCISE 2.1.1.   **Symmetry Relationships**

The impulse responses of real linear-phase filters can be either symmetric or antisymmetric and the point of symmetry can be either a sample value or the midpoint between two sample values. The point of symmetry depends on whether the filter length, N, is even or odd. As a result, there are four types of linear-phase FIR filters. These need to be treated separately for both filter design and implementation.

> Type   I: Symmetric impulse response, N odd.
> Type II: Symmetric impulse response, N even.
> Type III: Antisymmetric impulse response, N odd.
> Type IV: Antisymmetric impulse response, N even.

The four sequences given below are representatives of these four types. Use the *create file* option of **f siggen** to create filter files for these four sequences.

> I: $h[n] = -0.094\,\delta[n] + 0.303\,\delta[n - 1] + 0.600\,\delta[n - 2]$
> $\qquad +0.303\,\delta[n - 3] - 0.094\,\delta[n - 4].$
>
> II: $h[n] = -0.111\,\delta[n] + 0.062\,\delta[n - 1] + 0.512\,\delta[n - 2]$
> $\qquad +0.512\,\delta[n - 3] + 0.062\,\delta[n - 4] - 0.111\,\delta[n - 5].$
>
> III: $h[n] = -0.096\,\delta[n] + 0.283\,\delta[n - 1] + 0.0\,\delta[n - 2]$
> $\qquad -0.283\,\delta[n - 3] + 0.096\,\delta[n - 4].$
>
> IV: $h[n] = 0.010\,\delta[n] - 0.038\,\delta[n - 1] + 0.398\,\delta[n - 2]$
> $\qquad -0.398\,\delta[n - 3] + 0.038\,\delta[n - 4] - 0.010\,\delta[n - 5].$

(a) Linear-phase filters can be identified in the z-plane by looking at the locations of their zeros. Look at the zeros of each of these filters using **g polezero** and sketch the plots. Verify that for each zero not on the unit circle there is also a root at the conjugate reciprocal of that zero location. This means that the system function satisfies the relation

$$H(z) = H^*(1/z^*),$$

where the asterisk ($*$) denotes complex conjugation.

(b) Show analytically that the center sample of a Type III FIR filter must always be equal to zero.

(c) Which types of linear-phase filters always have a zero at $z = 1$? Which types always have a zero at $z = -1$? Show analytically that this will be true for all filters of a given type. To do this, write formulas for $H(z)$ evaluated at $z = 1$ and $z = -1$ in terms of the samples of the impulse response $h[n]$. Then use these formulas to show that Type III filters must always have a zero at $z = 1$. Summarize your results in your write-up.

(d) The first backward difference operator is defined by the difference equation

$$y[n] = x[n] - x[n - 1],$$

where $x[n]$ is the input and $y[n]$ is the output. Use **f lshift** and **f subtract** to take the first backward difference of each of the impulse responses and compare the roots of $y[n]$ with the roots of $x[n]$ for each of the signals that you generated above. What type of filter is the first backward difference of a Type I filter? What about the first backward difference of Type II, Type III, and Type IV filters?

(e) The inverse of the first backward difference operator is defined by the difference equation

$$y[n] = y[n - 1] + x[n],$$

where $x[n]$ and $y[n]$ are the input and output, respectively. This system is sometimes called an *accumulator*. Use **f lccde** to evaluate 15 samples of the output sequences for each of the four sequences that you generated. Notice that N will equal 1, M will be 0, $a_1 = -1$, $b_0 = 1$, and the initial conditions will be 0. Display each of the four outputs, using **g view**. For which symmetric signals does this operator produce an output signal of finite length? Which classes of output signals result from each of the different classes of input signals?

(f) Generalize the results of the previous two parts of this problem by finding twelve nontrivial operators that will map each class of linear phase sequences into each of the other classes. Omit the trivial operators that will map a Type I filter into a Type I filter, a Type II filter into a Type II filter, etc.   □

EXERCISE 2.1.2.   Magnitude and Phase Distortion

Both the magnitude and phase response characteristics of a filter can have an important effect on the filtered output. If the filter magnitude response is flat in the passband, the filter does not distort the input signal's spectral magnitude characteristics (in that spectral region). Similarly, if the phase response is linear, the filter does not distort the input signal's phase characteristics.

In this exercise, we examine the effects of magnitude and phase distortion in a couple of filtering examples. Use the *triangular wave* option in **f siggen** to generate the input signal $x[n]$, which should consist of four periods of a triangular wave each with a period of 32, an amplitude of 1, and a starting point of 0. Next, use the *sine wave* option of this same function to design two sine wave signals of length 64: the first with $\alpha = 1.0472$; and the second with $\alpha = 2.0944$. In both cases, let the amplitude be 1, the phase be 0, and the starting point be 0. Add the sine waves together using **f add** to form your second input signal, $v[n]$. Display $x[n]$ and $v[n]$ using **g view2** and sketch them.

(a) Two filters may have the same magnitude response but different phases. Two such filters are $h_0[n]$ and $h_1[n]$ contained in the files **f100** and **f101**, both of which are provided on disk. Use **g dtft** to plot the magnitude and phase responses for each and sketch these plots in your write-up. Notice that the magnitude responses are the same for both but that the phase responses are different.

(b) Two filters may also have the same phase response but different magnitude responses. Examine the magnitude and phase of the filter $h_2[n]$, which may be found in file **f102**, using **g dtft** and sketch the plots. Compare its characteristics to those of $h_0[n]$, and observe that the phase responses are identical but that the magnitude responses are different.

(c) Now perform the following filtering operations

$$y_0[n] = x[n] * h_0[n],$$
$$y_1[n] = x[n] * h_1[n],$$
$$y_2[n] = x[n] * h_2[n],$$
$$y_3[n] = v[n] * h_0[n],$$
$$y_4[n] = v[n] * h_1[n],$$
$$y_5[n] = v[n] * h_2[n],$$

using **f convolve**. Display and sketch $y_0[n]$ and $y_1[n]$ using **g view2**. The difference between these two signals is due to the difference in the phase response characteristics of the filters.

Display and sketch $y_0[n]$ and $y_2[n]$ using **g view2**. The difference now is in the magnitude responses of the filters. Both filters have the same phase response. For this example, which characteristic seems to have the greater impact on the output signal appearance, magnitude or phase?

(d) Now display and sketch $y_3[n]$ and $y_4[n]$ and also $y_3[n]$ and $y_5[n]$, again using **g view2**. The difference in signal appearances for the case of $y_3[n]$ and $y_4[n]$ is due to phase differences in the filters while for $y_3[n]$ and $y_5[n]$ it is due to magnitude differences. Which seems to have the greater impact on the output signal appearance in this case, magnitude or phase?

(e) In general, the relative importance of magnitude and phase distortion depends on the application and the signals involved. Think of a practical example in which the introduction of phase distortion would be very detrimental. Think of an example that would be very sensitive to magnitude distortion.   □

EXERCISE 2.1.3.   **Phase Response**

Use the *rectangular window* option in **f fdesign** to design an 18-point linear-phase lowpass filter with cutoff frequency $3\pi/4 = 2.3562$.

(a) Display the frequency response of this filter using **g dtft** and sketch $H_R(e^{j\omega})$, $H_I(e^{j\omega})$, and $|H(e^{j\omega})|$. At what frequencies does $H(e^{j\omega})$ have zeros on the unit circle?

(b) The phase response of a filter is generally evaluated by using the formula

$$\arg H(e^{j\omega}) = \arctan \frac{H_I(e^{j\omega})}{H_R(e^{j\omega})} + \begin{cases} \pi, & \text{if } H_R(e^{j\omega}) < 0, H_I(e^{j\omega}) > 0 \\ 0, & \text{if } H_R(e^{j\omega}) \geq 0 \\ -\pi, & \text{if } H_R(e^{j\omega}) < 0, H_I(e^{j\omega}) < 0. \end{cases}$$

Plot the phase response of this filter, using **g dtft**. Why isn't the phase plot a straight line?   □

EXERCISE 2.1.4.   Group Delay Calculation

The group delay response, $\tau(\omega)$, is defined as

$$\tau(\omega) = -\frac{d(\angle H(e^{j\omega}))}{d\omega},$$

where $\angle H(e^{j\omega})$ is the phase of $H(e^{j\omega})$. If $H(e^{j\omega}) \neq 0$, it can be evaluated using the equation

$$\tau(\omega) = \Re e\left[\frac{\mathcal{F}\{nh[n]\}}{H(e^{j\omega})}\right], \tag{2.2}$$

where $\mathcal{F}\{\cdot\}$ denotes the discrete-time Fourier transform and $\Re e[\cdot]$ is the real part operator.

(a) Derive this relation analytically.

(b) Write a macro to evaluate and plot the group delay of an FIR filter based on (2.2). Your macro should accept the input sequence $h[n]$ and a ramp sequence and produce a plot of the group delay. Your macro will likely use the functions **f fft**, **f multiply**, **f divide**, **f zeropad**, and **f realpart**. The plot can be displayed using **g look**.

(c) Using the *create signal* option in **f siggen**, create a file for the signal

$$h[n] = 0.4\delta[n] + \delta[n-1] + 0.4\delta[n-2].$$

The sequence length is three, the starting point is $n = 0$, the coefficients are real, and the three sequence values are 0.4, 1.0, and 0.4. Evaluate and plot its group delay using the macro. Compare the result of your macro with that produced by the function **g dtft**.

(d) (optional) Modify your macro to plot the group delay of an IIR filter. The simplest approach would be to use the instructions from the earlier macro along with the functions **f convert** and **f subtract**. Compare the performance of your macro with the function **g dtft** on the filter contained in the file **f003**.   □

EXERCISE 2.1.5.   Nonlinear-Phase

Consider the two simple first-order FIR systems with impulse responses given by

$$h_1[n] = \delta[n] - 0.9\,\delta[n-1]$$
$$h_2[n] = -0.9\,\delta[n] + \delta[n-1].$$

Use the *create file* option in **f siggen** to construct signal files for $h_1[n]$ and $h_2[n]$.

(a) Display and sketch the magnitudes of the frequency responses of these two systems, using **g dtft**.

(b) Display and sketch the phase and group delay responses of the two systems, using **g dtft**.

(c) Show that the cascade of these two systems is a linear-phase system by convolving $h_1[n]$ with $h_2[n]$, using **f convolve**. Examine and sketch the resulting impulse response, using **g view** and observe that it satisfies the linear-phase symmetry condition.

(d) Display and sketch the poles and zeros of $H_1(z)$ and $H_2(z)$, using **g polezero**. How is the zero of $H_1(z)$ related to the zero of $H_2(z)$?

(e) The system with impulse response

$$h[n] = \delta[n] + 3.4\,\delta[n-1] + 4.33\,\delta[n-2] + 2.448\,\delta[n-3] + 0.5184\,\delta[n-4]$$

is an example of an FIR system with a nonlinear-phase response. Using the observation that you made in part (d), find the impulse response, $g[n]$, of a linear-phase system that has exactly the same magnitude response and sketch it in your write-up. To do this, you should first create a signal file for $h[n]$, using **f siggen**. Then find the zeros of $h[n]$, using **g polezero**. Apply the appropriate options associated with that function to modify $h[n]$ and write it to an output file. Examine the group delay of your modified filter using **g dtft**. Is the group delay constant? If not, explain. □

EXERCISE 2.1.6.   **Heuristic Design**

The goal of this exercise is to design a linear-phase bandpass FIR filter by positioning zeros in the z-plane. The filter should roughly approximate the response

$$H(e^{j\omega}) \approx \begin{cases} 0, & 0 \le |\omega| \le 0.2\pi \\ 1, & 0.4\pi \le |\omega| \le 0.6\pi \\ 0, & 0.8\pi \le |\omega| \le \pi. \end{cases}$$

(a) Try to design a good *linear-phase* bandpass filter by placing up to 10 zeros in the z-plane using the *change pole-zero* option in **g polezero**. Begin the design by placing zeros at $z = \pm 1$, $z = 0.809 \pm j0.5878$, $z = -0.809 \pm j0.5878$. This initial placement of roots will help define the stopband. The remaining four zeros should be used to shape the passband. Notice that the linear-phase condition means that these four zeros must occur in conjugate reciprocal pairs.

    Attempt to make the magnitude response characteristics in the passband of the filter as flat as possible. Sketch the magnitude response and list the filter coefficients in your write-up. These coefficients can be displayed using the *list pole-zero* option in **g polezero**.

(b) Try to improve the magnitude response characteristics by placing the 10 zeros without the linear-phase constraint, but still with a real impulse response. Again, sketch the magnitude response and list the filter coefficients in your write-up. □

EXERCISE 2.1.7.   **Classical Smoothing Formulas**

The concept of FIR filters predates the computer age. A branch of the field of numerical analysis has long been concerned with developing formulas for interpolating, smoothing, differentiating, and integrating numerical data. Three classical smoothing formulas are given below.

$$y_1[n] = (x[n+1] + x[n] + x[n-1])/3.$$

$$y_2[n] = (x[n+2] + x[n+1] + x[n] + x[n-1] + x[n-2])/5.$$

$$y_3[n] = (-3x[n+2] + 12x[n+1] + 17x[n] + 12x[n-1] - 3x[n-2])/35.$$

These formulas are readily seen to correspond to linear-phase FIR lowpass filters. Create files for these filters using **f siggen** and determine and sketch their frequency responses using **g dtft**. Each of them should approximate a lowpass filter. Determine the approximate cutoff frequency of each.   □

2.2   FIR DESIGNS USING WINDOWS

The window method is one of the simplest and oldest methods available for designing FIR filters. It is also one of the most useful. Windows can be used to design FIR filters of any length. They can design filters with either linear or nonlinear-phase characteristics, and with either real or complex impulse responses. Furthermore, with the window design method, FIR filters can be designed quickly.

The window method is an example of a time-domain technique. It selects the impulse response of the filter, $h[n]$, to approximate an ideal impulse response, $i[n]$. Other filter design methods are frequency domain design algorithms. They select the filter's impulse response so that its frequency response approximates an ideal frequency response $I(e^{j\omega})$. This is not to say that the initial ideal filter specification cannot be given as a frequency response. But when the specification is given this way, an inverse DTFT must be performed first to evaluate $i[n]$.

The impulse response, $h[n]$, is formed by multiplying the ideal impulse response by an N-point window sequence, $w[n]$:

$$h[n] = i[n]w[n].$$

The resulting filter is of length N. When both the ideal impulse response and the window sequence are symmetric with the same point of symmetry, the filter will have linear phase.

Since the ideal impulse response is the product of two sequences, it follows from the multiplication property of the discrete-time Fourier transform that the frequency response of the filter will be the circular convolution of the ideal frequency response with the DTFT of the window,

$$H(e^{j\omega}) = \frac{1}{2\pi} \int_{-\pi}^{\pi} I(e^{j\theta}) W(e^{j(\omega-\theta)}) d\theta. \tag{2.3}$$

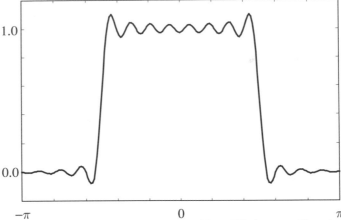

Figure 2.1. Amplitude response of a 32-pt FIR lowpass filter designed using a rectangular window.

A number of different sequences have been used as windows. The simplest of these is the *uniform* or *rectangular* window defined by

$$w[n] = \begin{cases} 1, & 0 \le n \le N - 1 \\ 0, & \text{otherwise.} \end{cases}$$

The discrete-time Fourier transform of the rectangular window is an aliased *sinc* function,

$$W(e^{j\omega}) = e^{-j\frac{N-1}{2}\omega} \frac{\sin\left(\frac{N\omega}{2}\right)}{\sin\left(\frac{\omega}{2}\right)}.$$

When this function is convolved with the ideal frequency response, $I(e^{j\omega})$, the resulting $H(e^{j\omega})$ exhibits Gibbs' oscillations at points of discontinuity in the ideal frequency response. These can be seen in Figure 2.1. Other choices for the window function can reduce these oscillations, but most do not eliminate them entirely.

Filters designed by the window method exhibit the following properties. Some of these are apparent in the example in Figure 2.1.

- The frequency response near a discontinuity of the ideal filter is approximately antisymmetric about a point of discontinuity in the neighborhood of that discontinuity.
- The width of the transition region is inversely proportional to the length of the window.
- The Gibbs' oscillations overshoot about equally on either side of the point of discontinuity, and they decay in amplitude as one moves away from the point of the discontinuity.

Changing the window function often reduces the size of the Gibbs' oscillations and produces a frequency response with smaller errors in the passbands and stopbands, but this

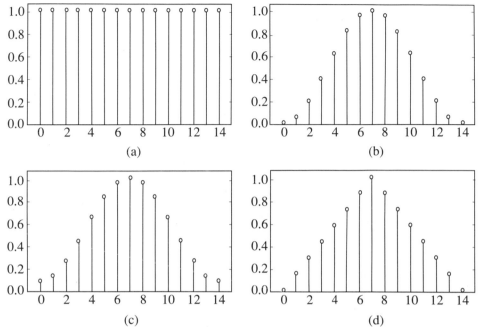

Figure 2.2. Four 15-pt window sequences: (a) Rectangular window, (b) Hanning window, (c) Hamming window, and (d) Bartlett window.

is almost always done at the expense of broadening the transition regions at discontinuities in the ideal frequency response. There are a number of other popular choices for window functions. These include the *hanning window* (or *von Hann window*),

$$w[n] = 0.5 + 0.5 \cos(\pi m[n]), \tag{2.4}$$

the *Hamming window*,

$$w[n] = 0.54 + 0.46 \cos(\pi m[n]), \tag{2.5}$$

and the *Bartlett* or *triangular window*,

$$w[n] = 1 - |m[n]|. \tag{2.6}$$

All of these have been defined in terms of the normalized time index $m[n]$, where

$$m[n] = \frac{n - \frac{N-1}{2}}{\frac{N-1}{2}}, \quad n = 0, 1, \dots, N - 1. \tag{2.7}$$

Clearly, $m[n]$ is confined to the range $-1 \leq m[n] \leq 1$ for $0 \leq n \leq N - 1$. Figure 2.2 illustrates these windows for $N = 15$. Their discrete-time Fourier transforms are shown in Figure 2.3. Some of these windows will be examined in the exercises that follow.

The DTFTs of these four windows are alike in their gross aspects. They differ, however, in the widths of the low frequency main lobes, which affect the filter transition

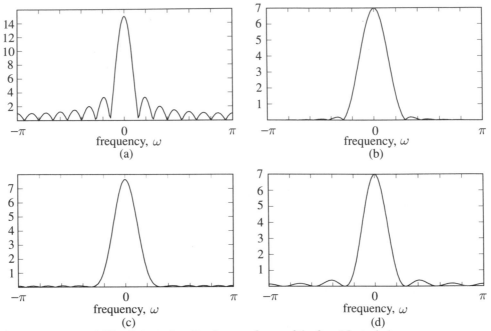

Figure 2.3. The discrete-time Fourier transforms of the four 15-pt window sequences from the previous figure: (a) Rectangular window, (b) Hanning window, (c) Hamming window, and (d) Bartlett window.

widths, and in the heights of their higher frequency sidelobes, which affect the size of the passband and stopband errors in the final design. The *Kaiser window* [1] is slightly different. It contains a parameter, α, that can be used to vary the tradeoff between the main lobe width and the sidelobe heights. It is given by

$$w[n] = \frac{I_0\left[\alpha\sqrt{1 - m[n]^2}\right]}{I_0[\alpha]},$$

where $I_0[\cdot]$ is the zero-order modified Bessel function of the first kind. In the limiting case when $\alpha = 0$, the Kaiser window reduces to a rectangular window. For a value $\alpha \approx 3.4$, it is very close to a Hamming window.

The value of α is chosen to meet specified constraints on the maximum errors in the passband and stopband of the filter. Once it has been chosen, the transition band specification can be met by varying the length, N, of the filter. Since the passband and stopband ripples of a filter designed by the window method are approximately equal, only one degree of freedom is available with respect to the errors in the frequency bands. Therefore, let $\delta = \delta_p = \delta_s$ denote the stopband and passband ripple and $A = 20\log_{10}(1/\delta)$ denote for the attenuation. Kaiser [1] determined empirically that the value of α required to achieve a particular value of A is approximately given by the formula

$$\alpha = \begin{cases} 0.1102(A - 8.7), & A > 50 \\ 0.5842(A - 21)^{0.4} + 0.07886(A - 21), & 21 \le A \le 50 \\ 0.0, & A < 21. \end{cases}$$

He also experimentally determined that the tradeoff among the filter length, ripple amplitude, and transition band width for frequency selective filters is governed by the relation

$$N \approx \frac{-20\log_{10}(\delta) - 7.95}{2.29\Delta\omega} + 1. \tag{2.8}$$

In this expression, δ is the peak ripple value and $\Delta\omega$ is the width of the transition band at each discontinuity measured in radians.

EXERCISE 2.2.1.   Comparing Some Classical Windows

Different choices for the window function result in filters with different transition bands and with different peak magnitude errors in the passband and stopband. This exercise quantifies some of those differences.

(a) Design a 45-point FIR lowpass filter using a rectangular window with a cutoff frequency of $0.25\pi = 0.7854$. To do this, simply use **f ideallp** to generate the filter. Since the rectangular window is an array of ones, nothing more needs to be done. Display and sketch the frequency response of the filter using **g dfilspec**. Measure the peak value of the passband frequency response and assume that this is $1 + \delta_p$. Find the highest frequency for which the magnitude response is greater than $1 - \delta_p$ and call this the passband cutoff frequency, ω_p. In a similar fashion, determine the peak stopband error, δ_s, and find the stopband cutoff frequency ω_s. Now record the passband ripple, δ_p, the stopband ripple δ_s, and the width of the transition band $\Delta\omega = \omega_s - \omega_p$.

(b) Repeat part (a) for a 45-point FIR filter designed with a hanning window, using the functions **f ideallp**, **f hanning**, and **f multiply**.

(c) Repeat part (b), using a Hamming window. Here you will use **f hamming** instead of **f hanning**.

(d) Repeat part (b), using a Blackman window. Here you will use **f blackman** instead of **f hanning**.

(e) Try repeating part (b) for a Bartlett window, using the function **f bartlett**. Describe the problem encountered in determining δ_p and δ_s. The Bartlett window is not a good window for filter design, in general.

(f) Summarize your observations. In particular, which windows seem to work well for designing filters? Which filter has the narrowest transition region? Which has the highest attenuation? Which has the fastest roll-off in the stopband?   □

EXERCISE 2.2.2. **Using Kaiser's Formula**

Kaiser's formula, which was given in (2.8), can be used to predict the filter order required to satisfy a given set of filter specifications when the filter is an FIR lowpass designed using a Kaiser window. This exercise should be completed after Exercise 2.2.1 has been done.

(a) Solve (2.8) for the ripple $\delta = \delta_p = \delta_s$ in terms of N and $\Delta\omega$.

(b) Solve (2.8) for the transition width, $\Delta\omega$, in terms of N and δ.

(c) The function **f eformulas** evaluates these estimation equations for you. Using parameters that you measured for filters designed using the rectangular, hanning, Hamming, and Bartlett windows in the previous exercise, calculate the ripple, δ, for the comparable filters designed using the Kaiser window. Compare those values with the values that you actually measured. Do these different windows appear to be better than, comparable to, or worse than the Kaiser window designs? □

EXERCISE 2.2.3. **Variation of Transition Width with Filter Length**

(a) Design a 16-point, 24-point, 32-point, and 48-point linear-phase FIR lowpass filter with a cutoff frequency of $\omega_c = \pi/3 = 1.0472$, using the window design method with a Hamming window. This can be done by using the **f ideallp**, **f hamming**, and **f multiply** functions. Display their magnitude responses by using **g dfilspec** and measure the width of their transition bands. (One way to do this is discussed in part (a) of Exercise 2.2.1.) Prepare a graph of the *reciprocal* of the transition width $(1/\Delta\omega)$ versus the filter length N.

(b) Repeat part (a) for a rectangular window. Plot your results superimposed on the plot for the Hamming window. Formulate some conclusions from these plots and include them in your write-up. □

EXERCISE 2.2.4. **Highpass Design**

(a) If $h_{LPF}[n]$ is a lowpass filter, then $h_{HPF}[n] = (-1)^n h_{LPF}[n]$ is a highpass filter. How are the cutoff frequencies of the lowpass and highpass filters related? Write a macro to compute the ideal impulse response of a causal, linear-phase highpass filter of length N and cutoff frequency $\omega_c = \pi/3 = 1.0472$ when N is an odd number. Your macro will likely make use of the functions **f ideallp** and **f cexp**. When the function **f ideallp** asks you for the cutoff frequency, remember that it is asking for the cutoff frequency of the *lowpass* filter. Verify the performance of your macro by displaying the magnitude response of several odd-length highpass filters, using **g dtft**.

(b) Design a 19-point linear-phase FIR highpass filter with a cutoff frequency $\omega_c = \pi/3 = 1.0472$, using a hanning window. Display and sketch the magnitude response of your filter. □

EXERCISE 2.2.5. **Bandpass Design to Meet Specifications**

In this problem you are to use a Kaiser window to design an odd-length FIR linear-phase bandpass filter of minimum length to meet the following specifications:

$$0 < |H(e^{j\omega})| < 0.01, \quad \text{for } 0 \le |\omega| \le 0.577,$$
$$0.99 < |H(e^{j\omega})| < 1.01, \quad \text{for } 0.994 \le |\omega| \le 2.148,$$
$$0 < |H(e^{j\omega})| < 0.01, \quad \text{for } 2.560 \le |\omega| \le \pi.$$

The functions **f eformulas** (option 1) and **f kalpha** (option 3) may be used to estimate the filter order and value of α, respectively. The first step in designing the filter is to obtain an ideal prototype filter. Do this by designing a lowpass filter using **f ideallp** with cutoff frequency $\omega_0 = 0.7854$. Use the **f cexp** function with $\omega_0 = 1.570796$ to modulate the lowpass prototype to $\pi/2$. Take the real part of the resulting complex sequence using the function **f realpart**. This will produce a cosine modulated bandpass prototype with passband amplitude of 1/2. Use the **f gain** function to scale the amplitude by a factor of 2.

Next, design the window using **f kaiser** and the value of α that you found earlier. Use the function **f multiply** to window the ideal bandpass prototype. Use **g dfilspec** to measure the frequency response of your filter and determine whether it satisfies the specifications. Since the formulas used to determine the design parameters only produce estimates, you may need to redesign the filter with an increased length. Remember that you are being asked for an odd-length filter. Therefore, the length should be increased by 2 in the next trial. After you have verified that the design meets the specifications, record the actual cutoffs and ripples for your filter. Sketch the magnitude response in your write-up. □

EXERCISE 2.2.6. **Hanning Window**

The Kaiser window can closely approximate a hanning window if its parameter α is appropriately chosen.

(a) Using the functions **f ideallp** and **f hanning**, design a 45-point hanning window lowpass filter with a cutoff frequency of $\omega_c = 0.4\pi = 1.2566$. Display and sketch its magnitude response, using **g dfilspec**, and measure the width of its transition band and the passband and stopband ripples.

(b) By designing several 45-point lowpass filters with the same cutoff frequency as in part (a) using Kaiser windows with different values of α, determine a value for α so that the Kaiser window and the hanning window will produce lowpass filters with approximately the same ripple values and transition widths. Use **f kalpha** to obtain an initial estimate of α. Construct the Kaiser window by using **f kaiser** to design the window, and **f ideallp** to design the lowpass prototype. The function **f multiply** should be used to apply the window to the prototype. Measure and record the transition width and the passband and stopband ripples for the best matching Kaiser filter. Which is the better filter, the Kaiser or the hanning? □

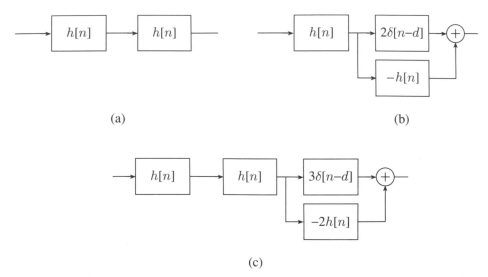

Figure 2.4. Some twicing configurations based on a single linear-phase FIR filter $h[n]$.

EXERCISE 2.2.7. Variation of Ripple Height with N

(a) Design four Hamming window lowpass filters, with lengths $N = 16, 32, 48, 64$, each with a cutoff frequency of $\omega_c = 0.4\pi = 1.2566$. Do this using **f hamming, f ideallp**, and **f multiply**. Examine the filters carefully, using **g dfilspec**, and record the ripple values, δ_p, δ_s, and stopband and passband cutoff frequencies.

(b) Plot the maximum passband and stopband ripples (i.e., $\max(\delta_p, \delta_s)$) as a function of N.

(c) Plot the transition widths as functions of the filter length N.

(d) Repeat parts (a) and (b) for the Blackman window. □

EXERCISE 2.2.8. Twicing

Kaiser and Hamming [2], based on some work of Tukey [3], proposed a technique, which they called *twicing*, that permits several copies of a filter to be combined to produce an improved response. We will explore this technique in this problem.

(a) Design an FIR filter of minimum odd length with a passband cutoff frequency of $0.4\pi = 1.2566$, stopband cutoff frequency of $0.6\pi = 1.8850$, and maximum ripple values of $\delta = 0.1$ in both the passband and stopband. Use the Kaiser window for the design, using the functions **f ideallp, f kalpha, f kaiser**, and **f multiply**. The function **f eformulas** can be used to obtain estimates of the minimum length. Record this length in your write-up. Call the resulting filter $h[n]$ and its frequency response $H(e^{j\omega})$.

(b) Compute the cascade of $h[n]$ with itself, i.e., $h[n] * h[n]$, using **f convolve**. Display and sketch the resulting frequency response, using **g dfilspec**, and measure the passband and stopband ripples of the cascade. How do these compare with those of $H(e^{j\omega})$? Determine the passband and stopband cutoff frequencies of the cascade and compare these with the original filter. Observe that the passband ripple is large compared to the stopband ripple.

(c) Now measure the frequency response of the arrangement shown in Figure 2.4b. The box labeled $2\delta[n - d]$ represents a delay of d samples with a gain of 2, where d is the delay of the filter $h[n]$ (i.e., if $h[n]$ is of length N, then $d = \frac{N-1}{2}$). The functions **f siggen**, **f gain**, and **f lshift** may be used to implement $2\delta[n - d]$. Measure the passband and stopband ripples and the passband and stopband cutoff frequencies as you did in part (b). Observe that this configuration produces a filter with a small passband ripple and a large stopband ripple.

(d) Repeat part (c) for the configuration shown in Figure 2.4c. Observe in this case that the resulting filter has approximately the same ripple in the passband and the stopband.

(e) Determine the total number of multiplications and additions required to implement the configuration in Figure 2.4c and compare these numbers with the number of operations required to implement a single Kaiser window filter with the same cutoff frequencies and attenuation. To do this, use **f eformulas** to determine the length of the equivalent Kaiser window filter. This illustrates that there is some loss in filter quality when twicing configurations are used.

(f) (optional) If the filters in part (d) are shifted so that they are zero-phase filters, the overall frequency response is $G(e^{j\omega}) = H(e^{j\omega})^2(3 - 2H(e^{j\omega}))$. The derivative of this polynomial taken with respect to $H(e^{j\omega})$ has a root at $H(e^{j\omega}) = 1$ and another at $H(e^{j\omega}) = 0$, the nominal passband and stopband gains. Extend this approach by finding a fifth-order polynomial in $H(e^{j\omega})$ whose derivative has double zeros at $H(e^{j\omega}) = 1$ and $H(e^{j\omega}) = 0$ and use that polynomial as the basis for a configuration that behaves like a lowpass filter with a nominal passband gain of one using five copies of the filter $H(e^{j\omega})$ (plus some delays and integer gains). Measure the passband and stopband ripples of the overall system using **g dfilspec**. They should be considerably smaller than those measured in part (d).   □

EXERCISE 2.2.9. Generation of Length Formulas for Hilbert Transformers

Comment. This exercise is really a small project and, if done properly, might be somewhat time consuming.

Develop a formula, similar to (2.8), which was developed for lowpass filters, that will relate filter length, transition width, and ripple height for FIR Hilbert transformers designed using a Kaiser window. To do this you will first need to design several Hilbert transformers of different lengths with different values of the Kaiser window parameter α using the function **f hilbert** to generate the impulse response of an ideal Hilbert transformer, the function **f kaiser** to generate the Kaiser windows, and the function **g dfilspec** to calculate and measure the frequency response.   □

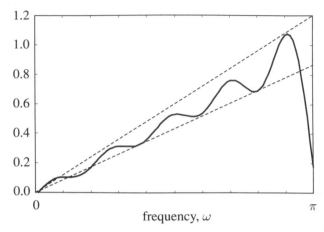

Figure 2.5. The magnitude response of a differentiator.

EXERCISE 2.2.10.   **Generation of Length Formulas for Differentiators**

Comment. This exercise is really a small project and, if done properly, might be somewhat time consuming.

Ripple heights for differentiators, and other filters whose passbands are not constant, increase as the magnitude response increases, as shown in Figure **??**. A useful measure of passband quality for such filters is the relative error, defined as the maximum value of the ratio of the magnitude error to the magnitude itself. For a differentiator, this can be computed from the slopes of the lines that bound the magnitude response above and below, as shown on that figure.

(a) In a manner similar to that suggested for the previous exercise, develop a formula, similar to (**??**), that will relate filter length and relative error for FIR differentiators of *even* length. Differentiators of even length can be designed to cover the whole frequency axis. Therefore, it is not necessary to include a transition band for these filters. This task will require the function **f diff** to generate the impulse response of an ideal differentiator, the function **f kaiser** to generate the Kaiser windows, and the function **g dfilspec** to calculate and measure the frequency responses.

(b) Repeat for differentiators of *odd* length. Here it will be necessary to include a transition band near $\omega = \pi$ and to include the transition width in the formula.   □

EXERCISE 2.2.11.   **Optimized Hanning Window**

The Hamming window is a generalization of the hanning window with reduced ripple. It is of the form:

$$w_h[n] = 0.54 - 0.46 \cos(\frac{2\pi n}{N-1}), \quad 0 \le n \le N - 1,$$

where N is the filter length. Notice that it is a weighted combination of a hanning window and a rectangular window. This exercise is concerned with the design of such an optimized window.

(a) Let $w_r[n]$ be an N-point rectangular window. Then our generalized window will be of the form

$$v[n] = \gamma w_r[n] - (1 - \gamma)\cos(\beta n)w_r[n].$$

Write a macro that will accept values for β, γ, and N, implement the window, and display the magnitude response of a resulting lowpass filter with a nominal cutoff frequency of $\pi/3 = 1.0472$. The function **f nlinear** can be used to generate the cosine function if a ramp function is specified as its input. The ramp can be generated using **f siggen** and the slope of the ramp can be adjusted using the function **f gain**.

(b) For a fixed window length of 19, vary the parameters β and γ to produce the window with the smallest stopband ripple. Record the value of the ripple and measure the transition width of the resulting filter. The function **g dfilspec** can be used to measure the stopband ripple.

(c) How does your window compare with the Hamming window?   □

EXERCISE 2.2.12.   **Determining the Ideal Prototype**

While it is straightforward to calculate the ideal impulse response to use with the window method analytically for many ideal frequency responses, when the ideal frequency response does not have a simple form, this step is often extremely difficult. A common method for approximating the ideal impulse response is to sample the ideal frequency response and then evaluate an inverse transform using the inverse DFT. This procedure is not perfect, as it can introduce temporal aliasing into the impulse response. In this exercise, you will explore the limitations of this approach.

(a) Design a 32-point FIR lowpass filter using a hanning window (**f hanning**) and the ideal impulse response

$$i[n] = \frac{\sin(0.5\pi(n - \frac{31}{2}))}{\pi(n - \frac{31}{2})}.$$

This ideal response is the one produced by the function **f ideallp**. Display and sketch the magnitude response of your filter using **g dfilspec**. Measure the width of the transition band of your filter and measure the passband and stopband ripple heights.

(b) Repeat your design using instead the ideal impulse response

$$i[n] = \frac{\sin(0.5\pi(n - \frac{31}{2}))}{32\sin(\pi(n - \frac{31}{2}/32))}.$$

To do this, generate the DFT of $i[n]$ using the following procedure. First generate a 32-point complex DFT sequence $I[k]$ using the *create file* option of **f siggen**.

$$I[k] = \begin{cases} (-1)^k, & 0 \le k \le 7 \\ 0.5 & k = 8 \\ 0 & 9 \le k \le 23 \\ 0.5 & k = 24 \\ (-1)^k, & 25 \le k \le 31. \end{cases}$$

Display $I[k]$ using **g view** and verify that the sample values are correct. Then calculate an inverse DFT using **f ifft** and take the real part of the result using **f realpart**. Multiply by the hanning window generated by **f hanning**. Display and sketch the magnitude response of your filter using **g dfilspec**. Measure the width of the transition band of your filter and measure its passband and stopband ripple heights.

(c) Repeat with the ideal impulse response

$$i[n] = \frac{\sin(0.5\pi(n - \frac{63}{2}))}{64 \sin(\pi(n - \frac{63}{2})/64)},$$

which results from using 64 samples of the ideal frequency response. Generate the sequence using a procedure that is similar to the one that you used in part (b). The resulting impulse response will contain 64 samples. Extract the 32 points for your filter by extracting the middle 32 samples using **f extract**. As before, measure the transition width and ripple heights of your design and compare with the filters that you designed in parts (a) and (b). How can the frequency sampling approach for determining prototypes be used to approximate the true ideal prototype with arbitrary closeness?   □

EXERCISE 2.2.13.   **Multiband Filters**

There are many ways to design multiband filters. One is simply to use the window method with a multiband ideal filter. One limitation of this approach is that all of the transition bands will be of the same width and all the ripple heights will be the same if the passband gains are the same. Another approach connects bandpass filters in parallel or bandstop filters in cascade.

(a) Design a multiband filter to approximate the following magnitude specification with a maximum magnitude error of 0.01 in each band.

$$H(e^{j\omega}) = \begin{cases} 1.0, & 0 \le \omega \le 0.1\pi \\ 0.0, & 0.15\pi \le \omega \le 0.25\pi \\ 0.5, & 0.35\pi \le \omega \le 0.6\pi \\ 0.0, & 0.8\pi \le \omega \le \pi. \end{cases}$$

For this part of the exercise, use a single multiband prototype and a Kaiser window. Your filter should be of minimum length. Use the function **f kalpha**, which implements Kaiser's design formula in (**??**), to determine the value of α from the ripple specification. Then use **f eformulas**, which is based on (**??**), to estimate the filter length for a Kaiser window filter. For the transition width, you should use the smallest of the three transition bandwidths in the filter. Generate the ideal filter as differences of lowpass filters using **f ideallp** and use the function **f kaiser** to determine the window. Use **g dfilspec** to measure the frequency response of your filter and verify that it meets the specifications. It may be necessary to consider several values of N to ensure that your filter is of minimum length. Record the length of your filter. Determine the number of multiplications and additions to implement the filter.

(b) Repeat the design of part (a), but design your filter as the sum of a lowpass filter and a bandpass, using different windows for the lowpass and bandpass filters. You should observe that you have more flexibility in determining the specifications for the individual filters. Try to minimize the number of additions and multiplications in the implementation. Record the length of your filter.   □

2.3 FIR DESIGN BY FREQUENCY SAMPLING

The frequency sampling design technique designs a length N FIR filter by specifying the values of N equispaced samples of the frequency response over the range of frequencies $0 \leq \omega < 2\pi$. Since these correspond to the DFT of the filter impulse response, $h[n]$ is found by simply computing an inverse DFT. This approach is intuitively satisfying because (1) these N values are sufficient to completely specify the filter and (2) because they bear a direct relationship to the frequency response. There are some limitations to the method, however, which will be explored in this section. The behavior of a linear, time-invariant system to arbitrary inputs depends on the whole discrete-time Fourier transform. Although the complete filter is specified by its DFT values, we do not gain design control over the whole DTFT. The actual frequency response of the filter $H(e^{j\omega})$ depends upon the interpolation between the sample values, which is only implicitly controlled. In addition, while the magnitude of the desired frequency samples may be known, the application may permit some flexibility in the specification of the phase or in the behavior of the magnitude response in the transition regions. Both of these can have a strong effect on the frequency domain interpolation and the efficacy of the method.

EXERCISE 2.3.1. **Effect of Delay on Behavior**

(a) Design a length 32 lowpass filter with a nominal cutoff frequency of $\pi/2 = 1.5708$ rad by setting

$$H[k] = \begin{cases} 1, & k = 0, 1, \ldots, 7 \\ 0, & k = 8, 9, \ldots, 24 \\ 1, & k = 25, 26, \ldots, 31, \end{cases} \tag{2.9}$$

where $H[k]$ is the 32-point DFT of $h[n]$. Generate the sequence $H[k]$, using the *create file* option of **f siggen**. Determine $h[n]$ by taking the real part of the 32-point inverse DFT, using **f ifft** and **f realpart**. Display and sketch the impulse response and the frequency response of your filter using **g view** and **g dfilspec**. Measure the peak values of the passband and stopband ripple. Verify that your filter does pass through the proper values at the frequencies where it was specified. Is this filter a good lowpass filter? Explain.

(b) Design a second filter by replacing the DFT samples above with the values

$$H_m[k] = (-1)^k H[k].$$

Note that $H_m[k]$ can be constructed from $H[k]$, using the **f cexp** function with $\omega_0 = \pi$. Enter the letters "pi" for π when using **f cexp**. Sketch the impulse response of this filter and compare it with the result in part (a). Display and sketch the frequency response of your filter and again verify that it possesses the proper magnitude values at the design frequencies. Measure the peak passband and stopband ripple in this case. Compare these results with those from part (a) and notice the improvement. Save this filter for use in part (d).

(c) Both of these designs correspond to special cases of the more general specification

$$H_\ell[k] = e^{-j\frac{2\pi\ell k}{N}} H[k]$$

for $\ell = 0$ and $\ell = N/2$. Find the value of ℓ, which minimizes the passband ripple by designing several other filters in this family. You can generate the signals using **f cexp** prior to computing the inverse DFT. Is this the same value of ℓ that minimizes the stopband ripple? You should consider both integer and half-integer values for ℓ.

(d) Measure the transition width and the ripple heights of the filter that you produced in part (b) and compare them with the values predicted by the Kaiser window formula in (2.8) using **f eformulas**. What does this suggest about the relative merits of this technique in comparison to the window method?   □

EXERCISE 2.3.2.   Effect of Transition Samples on Behavior

With the window method for designing FIR filters, we saw that oscillations in the passbands and stopbands could be made smaller if the transition bands were made wider. The same principle is true for frequency sampling designs.

(a) Consider the family of length 32 frequency sampling filters whose DFT values are given by

$$F[k] = \begin{cases} (-1)^k, & k = 0, 1, \ldots, 7 \\ \beta, & k = 8 \\ 0, & k = 9, 10 \ldots, 23 \\ \beta, & k = 24 \\ (-1)^k, & k = 25, 26, \ldots, 31. \end{cases}$$

The parameter β represents the magnitude of two frequency samples that have been placed in narrow transition bands. Let $h_1[n]$ represent the filter with $\beta = 0$. Determine $h_1[n]$ using the following procedure: First, generate $H[k]$ defined in (2.9) in the previous problem using the *create file* option in **f siggen**. Then, form $F[k]$ in the equation above using **f cexp** with $\omega_0 = \pi$. This results in $F[k]$ with $\beta = 0$. Finally, take the inverse DFT using **f ifft**. This result is $h_1[n]$. Examine its magnitude response using **g dtft** and verify that it is a lowpass filter.

Design another filter, $h_2[n] = 0.0625 \cos \pi n/2$, using **f siggen** with length 32 and starting point zero. This filter will be used to control the value of β in the experiment.

Write a two-line macro to implement the equation

$$h[n] = h_1[n] + \beta h_2[n].$$

It should use the functions **f add** and **f gain**. Systematically experiment with different values of β and examine the passband and stopband ripple heights using **g dfilspec**. Determine the value of β that minimizes the passband ripple height. Is this the same as the value that minimizes the stopband ripple height?

(b) Using the value $\beta = 0.5$ measure the ripple heights and the transition width and compare these results with those obtainable with a Kaiser window filter of the same order. Use the function **f eformulas** to estimate Kaiser window filter specifications for your comparison. List the advantages of each method.   □

2.4 EQUIRIPPLE DESIGNS

The window method and the frequency sampling method for FIR filter design both attack the design problem indirectly. Plausible strategies are used to select the filter coefficients that produce a reasonable frequency response. The quality of the resulting design is evaluated by measuring the width of the transition bands and the sizes of the errors (ripples) in the passbands and stopbands of the filter. An alternative is to approach the problem directly as an optimization problem in which the impulse response coefficients are regarded as free parameters whose values are adjusted to optimize a filter quality measure. There are two issues that must be addressed. First, transition width, stopband ripple, and passband ripple are separate measures of quality, each of which must be minimized. Thus the design problem must be recast to reflect a single quality measure that embodies all of them. Second, the computational complexity of general optimization algorithms increases rapidly with the length of the filter. If we wish to design large filters, we must use an efficient optimization procedure.

These issues are not trivial. In reading the historical papers leading up to the Parks–McClellan algorithm, you will see that several alternatives were tried before settling on the approach that has since become widespread. First, the class of filters that could be designed was restricted to the design of linear-phase Type I (odd length, symmetric) FIR filters. Later techniques were developed to design the other types of linear phase filters by solving an equivalent Type I design problem, but the technique is still limited to linear phase filters. By way of contrast both the windowing and frequency sampling methods can design FIR filters that are not restricted to have linear phase.

The Parks–McClellan algorithm [4,5] requires that the designer specify the length of the filter N, all of the passband and stopband cutoff frequencies, and the ratios between the values of the peak desired passband and stopband errors. The algorithm then minimizes the heights of the ripples simultaneously subject to the specified ratios. Formally, the frequency response of a length $2M + 1$ zero-phase ($h[n] = h[-n]$) FIR filter can be expressed as

$$H(e^{j\omega}) \; = \; \sum_{n=-M}^{M} h[n]e^{-j\omega n}$$

$$= h[0] + 2 \sum_{n=1}^{M} h[n] \cos(\omega n). \qquad (2.10)$$

The Parks–McClellan algorithm finds the unique set of filter coefficients that minimize the weighted Chebyshev error

$$E_\infty = \max_{\tilde{\omega}} \left| W(e^{j\omega}) \left\{ D(e^{j\omega}) - H(e^{j\omega}) \right\} \right|. \qquad (2.11)$$

Ideally, the set $\tilde{\omega}$ is the union of the passbands and stopbands of the filter. In practice, it is a dense set of frequency samples taken from the passbands and stopbands that includes the band edges. The function $D(e^{j\omega})$ is the (real) desired frequency response and $W(e^{j\omega})$ is a positive weighting function.

These filters are called *equiripple*, because when the optimal filter has been found, the frequency error function $E(e^{j\omega}) = W(e^{j\omega}) \left\{ D(e^{j\omega}) - H(e^{j\omega}) \right\}$ will take on its maximum value many times. Thus, the ripples in the filter are all of equal height. It can be proven, using the *alternation theorem*, that the error function must exhibit at least $M+2$ alternations in $\tilde{\omega}$. This means that there will be $M+2$ distinct points $\omega_1 < \omega_2 < \ldots < \omega_{M+2}$, where the error function takes on its maximum value

$$\left| E(e^{j\omega_i}) \right| = E_\infty \qquad i = 1, 2, \ldots, M+2, \qquad (2.12)$$

and where the weighted error alternates sign at each successive point

$$E(e^{j\omega_{i+1}}) = -E(e^{j\omega_i}) \qquad i = 1, 2, \ldots M+1. \qquad (2.13)$$

These points are called the *extremal frequencies*. There may be more than $M+2$ extremal frequencies; the theorem only sets a minimum.

There are a number of additional properties that can be proven concerning filters that minimize the Chebyshev error. If the filter is a lowpass, the maximum number of extremal frequencies is $M+3$. Filters that possess $M+3$ extrema are called *extraripple* filters. Thus, Type I lowpass filters will all possess $M+2$ or $M+3$ extrema. Furthermore two of these extrema will occur at $\omega = \omega_p$ and $\omega = \omega_s$. Filters with more than two bands can possess additional extrema. It is also true that the transition band response for a lowpass filter must be monotonic. This is not true for bandpass filters, as one of the exercises will show.

The strategy behind the Parks–McClellan algorithm is to find the set of extremal frequencies, rather than the filter coefficients. Since $H(z)$ for an FIR filter is a polynomial, it is completely specified by a number of samples equal to its length, which is one greater than its degree. The alternation condition provides the value of the frequency response at the extremal frequencies once the passband ripple is known. The actual steps that are implemented are the following:

1. Guess a set of extremal frequencies $\{\omega_i\}$, $i = 1, 2, \ldots, M+2$.

2. Calculate the deviation (which is equal to the peak error at convergence) associated

with this set. Parks and McClellan found that this is given by the formula

$$\delta = \frac{\sum_{k=1}^{M+2} b_k D(e^{j\omega_k})}{\sum_{k=1}^{M+2} \frac{b_k(-1)^{k+1}}{W(\omega_k)}},$$

where

$$b_k = \prod_{\substack{i=1 \\ i \neq k}}^{M+2} \frac{1}{\cos \omega_k - \cos \omega_i}.$$

Use the deviation to specify the frequency response at the extremal frequencies.

3. Interpolate to find the complete frequency response, calculate the resulting error function, and identify the new extremal frequencies.

4. If the measured extrema are the same as the set used in step 1, stop and calculate the impulse response corresponding to the frequency response calculated in step 3. Otherwise, replace the $\{\omega_i\}$ with the extrema found in step 3 and repeat steps 2–4.

To take full advantage of this algorithm, the relationship among the parameters ω_p, ω_s, δ_p, δ_s, and N must be determined. A precise relationship is not known; however, there are empirically derived relationships that have been formulated. One such relationship is due to Herrmann, et al. [6] and is given by

$$N = 1 + 2\pi \frac{D_\infty(\delta_p, \delta_s)}{\omega_s - \omega_p} - f(\delta_p, \delta_s) \frac{\omega_s - \omega_p}{2\pi}, \qquad (2.14)$$

where

$$D_\infty(\delta_p, \delta_s) = [a_1(\log_{10} \delta_p)^2 + a_2 \log_{10} \delta_p + a_3] \log_{10} \delta_s$$
$$[a_4(\log_{10} \delta_p)^2 + a_5 \log_{10} \delta_p + a_6], \qquad (2.15)$$
$$f(\delta_p, \delta_s) = 11.012 + 0.512(\log_{10} \delta_p - \log_{10} \delta_s), \qquad (2.16)$$

and δ_s is assumed to be less than or equal to δ_p. The coefficients in the formula are given by $a_1 = 0.005309$, $a_2 = 0.07114$, $a_3 = -0.4761$, $a_4 = -0.00266$, $a_5 = -0.5941$, and $a_6 = -0.4278$. This relationship can be manipulated so that any of the five parameters ω_p, ω_s, δ_p, δ_s, and N can be expressed in terms of the others, thereby providing a useful set of design formulas.

Kaiser [1] introduced a simpler expression for Parks–McClellan filters of a form similar to that of (2.8). It is given by

$$N = \frac{-20 \log_{10} \sqrt{\delta_p \delta_s} - 13}{2.324 \Delta\omega} + 1, \qquad (2.17)$$

where $\Delta\omega$ is the transition width defined as $\omega_s - \omega_p$.

This formula and the one in (2.8) can be used to compare optimal equiripple Parks–McClellan filters to Kaiser window filters. If $\delta_p = \delta_s$, the ripples of the equiripple design are approximately 5 dB smaller for the same values of the transition width and filter length than for a Kaiser window design. An additional advantage of the Parks–McClellan design is that δ_s does not have to be identical to δ_p, as with the window design methods.

Both empirical formulas for Parks–McClellan filters as well as the formula for Kaiser window filters are implemented in software in the function **f eformulas**.

EXERCISE 2.4.1. Looking at Parks-McClellan Designs

(a) Use the function **f pksmcc** to design a length 23 linear-phase FIR lowpass filter with $\omega_p = 0.2\pi$, $\omega_s = 0.3\pi$, and a passband ripple that is ten times larger than the stopband ripple. (This means that the passband weight should be one-tenth the value of the stopband weight.)

Note that when using **pksmcc** you should enter f_pc= 0.2 for the passband cutoff frequency, and f_sc= 0.3 for the stopband cutoff frequency.

(i) Use **g dfilspec** to measure the passband and stopband ripples of your filter.

(ii) Identify and record the extremal frequencies in your design. Count the number of alternations and compare this value with the number predicted by the alternation theorem.

(b) Repeat part (a) for a 35-point linear-phase FIR lowpass filter. □

EXERCISE 2.4.2. Equiripple Designs for Type II Filters

The original Parks–McClellan algorithm worked only for Type I designs. Type II filters can be designed by converting these to Type I design problems. Recall that all Type II filters have a zero at $z = -1$. Therefore, the system function for a Type II filter can be written as

$$H_2(z) = (1 + z^{-1})H_1(z),$$

where $H_1(z)$ is a Type I filter. The design problem can then be stated as the problem of finding the set of filter coefficients that minimize the error

$$E_\infty = \max_{\tilde{\omega}} \left| W(e^{j\omega}) \left\{ D(e^{j\omega}) - (1 + e^{-j\omega})H_1(e^{j\omega}) \right\} \right|$$

$$= \max_{\tilde{\omega}} \left| W(e^{j\omega})(1 + e^{-j\omega}) \left\{ D(e^{j\omega})/(1 + e^{-j\omega}) - H_1(e^{j\omega}) \right\} \right|,$$

which is in the form of a Type I equiripple design problem with a different ideal function and a different frequency weighting.

(a) Let $h_1[n]$ denote the solution to the above Type I approximation problem with the modified error function. Determine a formula for $h_2[n]$, the solution to the Type II problem, in terms of $h_1[n]$.

(b) If $h_2[n]$ is a $2N$-point FIR lowpass filter, what is the minimum number of alternations that it can possess?

(c) Use the function **f pksmcc** to design a length 24 linear-phase FIR lowpass filter with $\omega_p = 0.2\pi$, $\omega_s = 0.3\pi$ and a passband ripple that is ten times larger than the stopband ripple.

(i) Use **g dfilspec** to measure the passband and stopband ripples of your filter.

(ii) Identify the extremal frequencies in your design. Count the number of alternations and compare this value with the number predicted by your result in part (b).

(d) Use **f pksmcc** to design length 22 and length 23 FIR lowpass filters that satisfy the same specifications as in part (c). Record the passband ripple for all three filters using **g dfilspec**. Does the filter quality appear to improve with increasing filter length? □

EXERCISE 2.4.3. **Equiripple Designs for Type III Filters**

Type III FIR filters have an odd length and an antisymmetric impulse response. The frequency responses of Type III FIR filters always possess a zero at $z = -1$ and another at $z = 1$.

(a) Using the approach of Exercise 2.4.2, show analytically that a Type III FIR design can be recast as a Type I design problem with an appropriate modification to the ideal response and the weighting function.

(b) If $h_3[n]$ is a $(2N + 1)$-point FIR Hilbert transformer (Type III), what is the minimum number of alternations that it can possess?

(c) Use the function **f pksmcc** to design linear-phase FIR Hilbert transformers with a lower cutoff frequency of $\omega_1 = 0.1\pi$ and an upper cutoff frequency of $\omega_2 = 0.9\pi$ for filter lengths of 15, 17, and 19. Identify the extremal frequencies in your designs. Count the number of alternations and compare this value with the number predicted by your result in part (b). You should observe that the 15-point filter and the 17-point filter have the same number of alternations. Explain why this is consistent with the alternation theorem. □

EXERCISE 2.4.4. **Equiripple Designs for Type IV Filters**

Type IV FIR filters have an even length and an antisymmetric impulse response. The frequency responses of Type IV FIR filters always possess a zero at $z = 1$.

(a) Using the approach of Exercise 2.4.2, show analytically that a Type IV FIR design can be recast as a Type I design problem with an appropriate modification to the ideal response and the weighting function.

(b) If $h_4[n]$ is a $2N$-point FIR Hilbert transformer (Type IV), what is the minimum number of alternations that it can possess?

(c) Use the function **f pksmcc** to design a length 14 linear-phase FIR Hilbert transformer with a lower cutoff frequency of $\omega_1 = 0.1\pi$, and an upper cutoff frequency of $\omega_2 = 0.9\pi$.

(i) Use **g dfilspec** to measure the ripple of your filter.

(ii) Identify the extremal frequencies in your design. Count the number of alternations and compare this value with the number predicted by your result in part (b). □

EXERCISE 2.4.5.   Empirical Formulas

In order to use the Parks–McClellan algorithm effectively to design filters that meet given frequency domain specifications, a relationship among the parameters ω_p, ω_s, δ_p, δ_s, and N must be determined. Two such relationships were given: one in (2.14) and one in (2.17). Both are empirically derived and thus only provide estimates. In this exercise, you will examine the accuracy of these formulas for a few examples and compare them.

(a) Design the following four equiripple lowpass filters each of length 30, using **f pksmcc**. Each filter has $\omega_p = 0.2\pi$ and $\omega_s = 0.3\pi$.

$$\begin{array}{lll} h_1[n]: & \text{passband weighting} = 1; & \text{stopband weighting} = 1 \\ h_2[n]: & \text{passband weighting} = 1; & \text{stopband weighting} = 2 \\ h_3[n]: & \text{passband weighting} = 1; & \text{stopband weighting} = 4 \\ h_1[n]: & \text{passband weighting} = 1; & \text{stopband weighting} = 8 \end{array}$$

Using **g dfilspec**, determine and record the actual values of δ_p and δ_s for each filter. For convenience you may wish to use the following macro called **design.bat**:

> **f pksmcc _file   4   2   30   0.2   1   0.3   %1   25**
> **g dfilspec   _file**

To design and evaluate a filter using the macro, type

> **design   w**

where **w** is the value of the stopband weighting (i.e., 1, 2, 4, or 8).

(b) Using the function **f eformulas** to evaluate the Parks–McClellan formula [see (2.14)] and Kaiser's formula for Parks–McClellan filters [see (2.17)], estimate the predicted length for each filter given the passband and stopband ripples measured in part (a). Note that the transition width is 0.31416. Which formula seems to be more accurate? Based on your observations, do these formulas provide reasonable parameter estimates?   □

EXERCISE 2.4.6.   Convergence

The Parks–McClellan procedure is an iterative procedure. This exercise will look at its rate of convergence. It is not recommended for computers that do not have a floating-point coprocessor.

(a) Consider the improvement in filter quality after successive iterations of the Parks–McClellan algorithm. Specifically, perform various numbers of iterations of the Parks–McClellan algorithm using **f pksmcc** for a 7-point FIR lowpass filter with passband cutoff frequency $\omega_p = 0.2\pi$, stopband cutoff frequency $\omega_s = 0.4\pi$, and $\delta_s = \delta_p$. In order to vary the number of iterations, you will have to run the function repeatedly for different values of the parameter that controls the maximum number of iterations. Save the filters after each run to help answer the questions in part (b). Use **g dfilspec** to measure the peak value of the error. You can do this by measuring the largest ripples in the passband and stopband. Does the peak error decrease monotonically? How many iterations are required for convergence?

(b) Produce a graph of the deviation as a function of the iteration number for the filters that you designed in part (a). The deviation is the quantity δ that is computed at the beginning of each iteration. You can measure the deviation at the end of each iteration because the magnitude response at the passband cutoff frequency is $1 - \delta$. According to the mathematics that supports the Parks–McClellan algorithm, the deviation should increase monotonically with the number of iterations. Are your plots consistent with this fact?

(c) Repeat (a) for a 17-point filter. Does the number of iterations required to converge appear to vary with the filter length?   □

EXERCISE 2.4.7.   A Conjecture

This exercise seeks an answer to the question of whether or not an equiripple filter can be designed using a window. Recall for a window design that the window function is chosen independently of the ideal impulse response and that once a window family has been selected, the window coefficients depend only on the window length.

(a) Design a 10-point lowpass filter with $\omega_p = 0.1\pi = 0.3142$, $\omega_s = 0.3\pi = 0.9425$, and with equal passband and stopband ripples using **f pksmcc**. Use **f ideallp** to determine the ideal impulse response of a 10-point ideal lowpass filter with a nominal cutoff frequency of $\omega_c = 0.2\pi = 0.6283$, and then, using **f divide**, calculate the coefficients of the effective window function. Examine the window using **g view** and sketch it.

(b) Repeat part (a) for a 10-point lowpass filter with $\omega_p = 0.4\pi = 1.2566$, and $\omega_s = 0.6\pi = 1.8850$. Are the two windows the same? The function **g sview2** can be used to display them simultaneously. Does this suggest that equiripple filters can be designed using an appropriate window?   □

EXERCISE 2.4.8.   Highpass FIR Filters

(a) Design an 18-point equiripple FIR highpass filter with stopband cutoff frequency $\omega_s = 0.3\pi$ and passband cutoff frequency $\omega_p = 0.5\pi$. Use **f pksmcc** for your design and make the desired passband ripple equal to the desired stopband ripple. Display and sketch the magnitude response of your filter, using **g dfilspec**. Explain why the magnitude response is distorted.

(b) Repeat part (a) for a 17-point filter. Why is your result better in this case?

(c) Repeat (a) and (b) for two lowpass filters. (Interchange the values for the passband and stopband cutoff frequencies.) Explain why the phenomenon observed in parts (a) and (b) does not occur for lowpass filters.   □

EXERCISE 2.4.9.   Transition Behavior in Bandpass Filters

In this exercise you investigate the transition band behavior of bandpass filters designed using the Parks–McClellan algorithm. Using **f pksmcc** with equal weights in each band, consider the design of a length 39 bandpass filter with

$$\begin{aligned}\text{stopband region} \quad & 0 \leq |\omega| \leq 0.1\pi, \\ \text{passband region} \quad & 0.4\pi \leq |\omega| \leq 0.6\pi, \\ \text{stopband region} \quad & 0.7\pi \leq |\omega| \leq \pi.\end{aligned}$$

The first step is to type **f pksmcc** and specify an output file name. Then select option (1), the multiple passband/stopband option, for the filter type desired. Since this is a bandpass filter with lower stopband, passband, and upper stopband regions, there are a total of three bands. Enter 3 for the number of bands. To design the filter to be of length 39, select option 2 (filter length) and enter 39 when prompted for the length.

Next the program asks you to specify each of the 3 bands in terms of a) being a passband or stopband, b) its lower and upper cutoff frequencies, and c) a value for the weighting in that band. For band 0, enter 0 to indicate that it is a stopband, enter f_lc= 0 to indicate that the stopband begins at 0 and f_uc= 0.1 to indicate that it ends at 0.1π, and 1 for the weighting function.

For band 1, enter 1 for passband, 0.4 for f_lc, 0.6 for f_uc, and 1 for the weighting. Finally, for band 2, enter 0 for stopband, 0.7 for f_lc, 1 for f_uc, and 1 for the weighting. Let the Remez exchange iterations run through 25 cycles. Observe that your entries result in two regions that are unspecified, i.e., don't care regions. Those are the regions between 0.1π and 0.4π, and 0.6π and 0.7π. Thus the Parks–McClellan algorithm will allow the filter to assume any spectral shape in these don't care regions as long as the error in the specified regions is minimized.

(a) Display and sketch the frequency response of your filter using **g dtft**.

(b) You should have observed that the filter has a large peak in the middle of the first transition band. Try to redesign the filter to remove this peak by modifying the band edge specifications of the filter. Sketch the frequency response of your final design and record the specifications you used.   □

EXERCISE 2.4.10.   Sparse Filters

The Parks–McClellan algorithm will find the linear-phase FIR filter of a given length that has the smallest passband and stopband ripples for a given set of cutoff frequencies. It is not guaranteed to find the optimal filter for a given number of nonzero coefficients.

(a) Use **f pksmcc** to design an 11-point FIR filter with equal passband and stopband weightings. Let the passband cutoff frequency be $\omega_p = 0.2\pi = 0.6283$ and the stopband cutoff frequency be $\omega_s = 0.4\pi = 1.2566$. Measure and record the peak ripple height for your filter using **g dfilspec**.

(b) Repeat the design in part (a) for a Kaiser window filter using the functions **f ideallp**, **f kaiser**, **f multiply**, and **f eformulas**. How does the performance of the Kaiser window design compare with the equiripple design?

(c) Now design another lowpass filter that has 11 nonzero coefficients, but which may have a longer total length than 11 samples, i.e., by spreading the nonzero samples out and inserting zeros. You could do this by looking at the ideal impulse response and retaining its 11 largest samples. You may wish to add a window function. Your design should retain its linear-phase characteristic, but may have a longer overall delay. Measure the peak ripple height on your filter and sketch its frequency response.   □

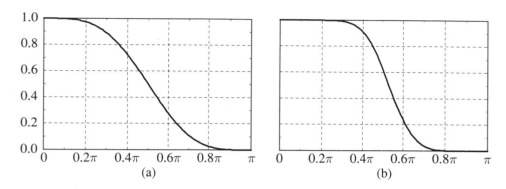

Figure 2.6. Two examples of maximally flat FIR filters. (a) $p = 1$, $q = 1$. (b) $p = 4$, $q = 5$.

2.5   MAXIMALLY FLAT DESIGNS

Hamming [7] has shown that a frequency response of the form

$$H(e^{j\omega}) = \frac{\int\limits_{-1}^{\cos\omega} (1 + t)^p (1 - t)^q dt}{\int\limits_{-1}^{1} (1 + t)^p (1 - t)^q dt} \tag{2.18}$$

corresponds to a zero-phase FIR lowpass filter with a maximally flat frequency response for integer values of $p \geq 0$ and $q \geq 0$. The impulse response for this filter can be obtained in stages. Begin by integrating the numerator of (2.18) to obtain a polynomial of degree M in $\cos\omega$. Then, normalize this by dividing by the value of the denominator integral, which is a constant. Finally the Chebyshev identity $\cos\omega n = T_n(\cos\omega n)$, where $T_n[x]$ is the nth Chebyshev polynomial, can be used to convert the polynomial into a linear combination of functions $\cos kn$ for $k = 0, 1, \ldots, M$, for which the inverse Fourier transform can be found trivially. An implementation of this procedure is provided in the function **f maxflat**.

The term *maximally flat* means that the frequency response is monotonically decreasing. The first p derivatives of the frequency response at $\omega = \pi$ and the first q derivatives at $\omega = 0$ are zero. Two examples of maximally flat FIR filters are shown in Figure 2.6.

The denominator of (2.18) simply represents a normalizing constant, and the numerator term corresponds to a polynomial in $\cos\omega$ of degree $p + q + 1$. A careful inspection of (2.18) and the inverse DTFT definition shows that the the length of maximally flat filters is twice the degree of the polynomial plus one. Thus, if the filter length is N, then $N = 2p + 2q + 3$.

Maximally flat filters are noncausal with zero phase but can be converted to causal, linear-phase filters by simply shifting the impulse response appropriately. The cutoff frequency of the filter is controlled indirectly through the choice of p and q. The inflection point, ω_I, of the frequency response occurs at approximately the frequency

$$\omega_I = \cos^{-1} \left(\frac{p-q}{p+q} \right). \tag{2.19}$$

EXERCISE 2.5.1. Maximally Flat Designs

(a) Find the filter corresponding to the case $p = q = 1$ using **f maxflat**. Display and sketch its frequency response using **g dfilspec** and determine the frequency at which $H(e^{j\omega}) = 1/2$.

(b) Repeat part (a) for the additional filters

    (i) $p = 2, q = 2$.

    (ii) $p = 3, q = 1$.

    (iii) $p = 6, q = 2$.

    (iv) $p = 2, q = 6$.

In each case, compare the nominal cutoff frequency with the inflection point predicted by (2.19). You should not expect to get exact agreement. □

EXERCISE 2.5.2. Adjusting the Cutoff Frequency

One problem with the maximally flat design procedure is that the designer does not have complete control over the cutoff frequency. One way to achieve this at the expense of some optimality is to design two maximally flat filters $H_1(e^{j\omega})$ and $H_2(e^{j\omega})$, where the inflection point of $H_1(e^{j\omega})$ is below the desired cutoff frequency and the inflection point of $H_2(e^{j\omega})$ is above it. We then set

$$h[n] = \lambda h_1[n] + (1 - \lambda)h_2[n]$$

for $0 \leq \lambda \leq 1$. Varying the value of λ then varies the inflection point of $H(e^{j\omega})$ between the inflection points of H_1 and H_2.

(a) Show analytically that the DTFT of $h[n]$ [i.e., $H(e^{j\omega})$] is monotonically decreasing.

(b) Assume that p_1 and q_1 are the filter parameters for $h_1[n]$, p_2 and q_2 are the parameters for $h_2[n]$, and that $p_1 = p_2 + 1$ and $q_1 = q_2$. Determine the length of $h[n]$. How many derivatives of $H(e^{j\omega})$ are zero at $\omega = 0$ and at $\omega = \pi$? These numbers give us a measure of how much optimality we have lost.

(c) We wish to design $h[n]$ to have a monotonic response with a transition band gain of $1/2$ at $\omega = 0.45\pi = 1.414$. Let $q_1 = q_2 = 1$, $p_1 = 2$, and $p_2 = 1$ and use the function **f maxflat** to design $h_1[n]$ and $h_2[n]$. Then use **g dfilspec** to measure $H_1(e^{j\omega})$ and $H_2(e^{j\omega})$ carefully at $\omega = 0.45\pi = 1.414$ and calculate the appropriate value of λ.

(d) Combine $h_1[n]$ and $h_2[n]$ using **f lshift**, **f gain**, and **f add** to produce $h[n]$. Be careful when you do this, since $h_1[n]$ and $h_2[n]$ are of different lengths. Both must be properly aligned in time so that $h[n]$ will have linear phase. Display and sketch $H(e^{j\omega})$ using **g dfilspec** and verify that it has the correct cutoff. □

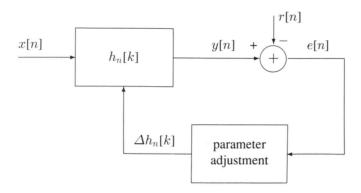

Figure 2.7. An adaptive FIR filter.

2.6 ADAPTIVE FIR FILTERS

Adaptive FIR filters' coefficients change with time as the filter input changes. Adaptive filters have been applied to problems of modeling unknown systems, compensation of communication channels, noise removal, and echo cancellation. Adaptive filters is an active area of research, and several texts have been written on the topic [8–11]. The sole goal of this section is to look at the simplest adaptive filter—an adaptive FIR filter whose coefficients are updated using the LMS algorithm.

Figure 2.7 shows an adaptive filter and its related computational elements. The filter itself is an FIR filter whose coefficients have the instantaneous values $h_n[k]$. Thus,

$$y[n] = \sum_{k=0}^{N-1} h_n[k]x[n-k].$$
(2.20)

The system has two inputs—the input to the filter, $x[n]$, and a reference input, $r[n]$. The filter coefficients are adjusted to drive the output $y[n]$ toward the reference signal $r[n]$. The error signal, $e[n]$, defined as

$$e[n] = y[n] - r[n],$$
(2.21)

measures the difference between the filter output and the reference signal.

The filter coefficients are adjusted to minimize the expected value of the square of the error energy

$$\mathcal{E} = E(e^2[n]).$$
(2.22)

This is done iteratively using the method of *steepest descent*. If $h_n[k]$ denotes the kth coefficient at time n, then

$$h_{n+1}[k] = h_n[k] - \mu \frac{\partial \mathcal{E}}{\partial h_n[k]}.$$
(2.23)

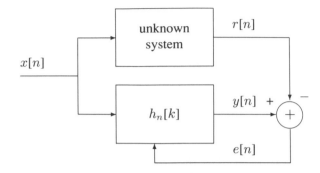

Figure 2.8. An adaptive filter used for system identification.

The parameter μ is known as the *step size*. It controls the rate of adaptation and the stability of the overall system.

From (2.20), (2.21), and (2.22),

$$\frac{\partial \mathcal{E}}{\partial h_n[k]} = 2E(e[n]x[n-k]). \tag{2.24}$$

The difficulty with this expression is the fact that the expectation is difficult to compute. Widrow and Hoff's solution [12] is to replace the expected value by the instantaneous value. This leads to the following set of equations, which is known as the *LMS algorithm*.

$$
\begin{aligned}
y[n] &= \sum_{k=0}^{N-1} h_n[k]x[n-k] \\
e[n] &= y[n] - r[n] \\
h_{n+1}[k] &= h_n[k] - 2\mu e[n]x[n-k] \quad k = 0, 1, \ldots, N-1.
\end{aligned}
$$

Depending upon the application, $y[n]$, $e[n]$, or $h_n[k]$ can be considered to be the system output. The LMS algorithm, as embodied in the above equations, is implemented by the function **f adaptfir**.

EXERCISE 2.6.1.   **System Identification**

One of the most basic uses of adaptive filters is for system identification. The filter is configured as shown in Figure 2.8. The same broadband input is applied to the adaptive filter and to an unknown system, and the coefficients of the adaptive filter are adjusted until the outputs of the two systems agree. After adaptation, the system function of the unknown system can be approximated by the system function of the adaptive processor. Adaptive system identification can be used to model a system whose parameters are slowly varying when the input and output signals are both available, for example in vibration studies of mechanical systems. In realistic situations, the output of the unknown system can be expected to be distorted by additive noise.

(a) For this first part of the problem, let's see whether the system will work for a simple example. Create the following signals and systems.

(i) Let the unknown system be an FIR filter with the following impulse response:

$$h[n] = \delta[n] + 1.8\delta[n-1] + 0.81\delta[n-2].$$

The *create file* option in **f siggen** can be used to create this filter. At the same time, create another filter file to contain the initial values of the adaptive filter coefficients, i.e. $h_n[k]$ when $n = 0$. Let the length be $N = 4$ and set the initial coefficient values to zero.

(ii) For the input signal $x[n]$ create 100 samples of a random sequence using the function **f rgen** with an initial seed of 1. Create the reference signal $r[n]$ by passing $x[n]$ through the filter using **f convolve**, then use **f truncate** to equalize the lengths of the input and reference signals.

(iii) Use **f adaptfir** to implement the filter. The function will ask you to specify the files containing the samples of $x[n]$, $h_0[k]$ (the initial coefficients of the filter), $h_M[k]$ (the final coefficients of the filter, where M equals the length of $x[n]$ and $r[n]$), $y[n]$, and $e[n]$.

Initially set the step size to $\mu = 0.5$, let the filter adapt, and record the final set of coefficients, $h_M[n]$.

(b) If the response of the adaptive filter is converging toward the response of the unknown, $e[n]$ should decay exponentially to zero as M increases. Display and sketch $e^2[n]$, using **f multiply** to calculate $e^2[n]$. How many iterations of the LMS algorithm were necessary for the mean squared error to fall to 10% of its peak value?

(c) Repeat for $\mu = 0.1$ and $\mu = 0.05$.

(d) Repeat for $N = 3$ and $N = 2$ with $\mu = 0.5$.

The adaptive filter will converge to the proper solution when the unknown system is actually an FIR filter. The rate of convergence depends on the step size. The larger the step size the faster the convergence. However, when noise is present, a smaller step size may converge to a more accurate result. This fact is explored in the next exercise.   □

EXERCISE 2.6.2.   **Effect of Measurement Error**

This exercise repeats the steps of Exercise 2.6.1, but with uniform random noise added to the reference signal. This can be done by replacing the reference signal by the signal

$$\tilde{r}[n] = r[n] + \gamma v[n].$$

Create the signal $v[n]$ using the function **f rgen** with a seed value of 2. This signal can be scaled using **f gain** and added to $r[n]$ using **f add**.

To measure the convergence of the adaptive filter define the following three errors:

$$E_1 = \sum_{n=0}^{3} \{h[n] - h_{1000}[n]\}^2$$

$$E_2 = \sum_{n=80}^{99} \{e[n]\}^2$$

$$E_3 = \sum_{n=980}^{999} \{e[n]\}^2 .$$

The first of these measures the actual errors in the coefficients ($h[n]$ is the actual impulse response of the unknown system.) The other errors estimate the average power in the error signal after 90 and 990 iterations. These quantities can be measured from the output signals of **f adaptfir** using **f extract**, **f subtract**, and **f summer**.

(a) Using the procedure described in the previous exercise calculate the error signal and final coefficient values for an adaptive filter with $N = 5$ coefficients. Let $h[n]$ be a five-point filter corresponding to the "unknown" system defined in Exercise 2.6.1(a). Let the length of $x[n]$ and $r[n]$ be 1000 samples rather than the 100 samples specified in that exercise. Use $\tilde{r}[n]$ as the reference signal instead of $r[n]$. Calculate the errors E_1, E_2 and E_3 for the four cases corresponding to γ equal to 0.5 and 0.1 and μ equal to 0.5 and 0.1. Arrange your measurements in a table.

(b) Which error is the more accurate measure of the convergence of the algorithm? Explain why the error signal, $e[n]$, is not accurately measuring convergence. Can you suggest alternative errors that might be more reliable?

(c) How does the modeling accuracy depend upon the step size μ? How does it depend upon the amount of measurement noise γ? □

EXERCISE 2.6.3. Modeling Error

An adaptive FIR filter will model any unknown system with an FIR model. In this problem, we want to see how well such a system can model an IIR system. The arrangement for the filter is the same as the one shown in Figure **??** that was used for the previous two exercises. Let the unknown system be an IIR filter with the system function

$$H(z) = \frac{1 + 0.5z^{-1}}{1 - 0.5z^{-1}}.$$

The function **f siggen** can be used to generate the filter. Generate a 300-point uniform random sequence with unit amplitude using the function **f rgen** with a seed value of 1. Use this signal for the input to both the unknown system and to the adaptive filter. The function **f filter** can be used to generate 300 samples of $r[n]$. Use **f adaptfir** to implement the system with $\mu = 0.3$ and filter length $N = 1, 2, 3, 4, 5$ with initial filter coefficient values of zero. Record the final values of the filter coefficients. Calculate the error

$$E_N = \sum_{n=280}^{299} e_N^2[n]$$

where e_N is the error for the filter of length N. This error can be computed using the functions **f extract** and **f summer.** Plot a curve of E_N versus N. What seems to be a "reasonable" value to use? How does this change as the location of the single pole in $H(z)$ is varied? □

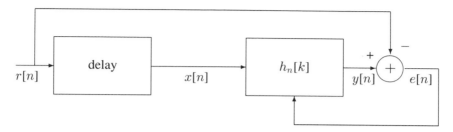

Figure 2.9.   An adaptive filter used for prediction.

EXERCISE 2.6.4.   **Signal Prediction**

The prediction application is also fairly straightforward. The input to the adaptive filter is a delayed version of the desired signal. To the extent that the adaptive filter can predict the future samples of the input process, the error signal will be driven to zero. Prediction is used for signal compression and noise reduction. The system is configured as in Figure 2.9.

A sinusoidal signal is an example of a signal that is predictable from its past values. Using trigonometric identities, we can show that

$$\sin \theta n = 2 \cos \theta \sin \theta (n - 1) - \sin \theta (n - 2).$$

Thus, if $r[n] = \sin \theta n$, then

$$r[n] = 2 \cos \theta r[n - 1] - r[n - 2].$$

For an input signal use **f siggen**, **f rgen**, **f gain**, and **f add** to generate 1024 samples of the signal

$$r[n] = \sin \alpha n + \epsilon[n], \quad 0 \leq n \leq 1023,$$

where $\alpha = 3\pi/16 = 0.5891$ and $\epsilon[n]$ is uniform white noise with a maximum amplitude of 0.02. Use 1 as an initial seed in **f rgen** to generate $\epsilon[n]$ and **f gain** with a gain factor of 0.04 to scale $\epsilon[n]$ to have a maximum amplitude of 0.02.

(a) Generate a zero-valued FIR filter of length two to be used as $h_0[n]$, the starting solution in the adaptive filter. Then, generate $x[n] = r[n + 1]$ using **f lshift** with a shift of -1. Run the adaptive filter using **f adaptfir** with $\mu = 0.03$ and determine the final set of filter coefficients. Record these. Also look at the error signal $e[n]$ using **g view** and observe that it gets smaller as n increases.

(b) Calculate the value of the filter coefficients that should result if there were no noise present in the input signal.

(c) In theory, a two-point adaptive FIR filter should be adequate to remove the sinusoidal component. What happens when a three-point filter is used? What about for a four-point filter? Look at both the filter coefficients and the error function, $e[n]$. Explain the differences you observe.   □

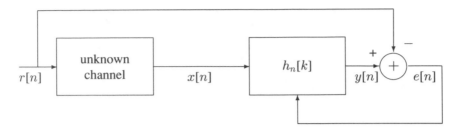

Figure 2.10. An adaptive filter used for compensation of communications channels.

EXERCISE 2.6.5.   Channel Compensation

Many communications channels can be viewed as random linear, slowly time-varying systems. Adaptive filters are often used at the receiver to compensate for the amplitude and phase distortion of these unknown channels that behave like LTI filters with unknown (and possibly time-varying) frequency (impulse) responses. The partially compensated channel allows data to be transmitted at a higher rate. An adaptive filter used for channel compensation is configured as shown in Figure **??**. During a *training interval*, a predetermined sequence is transmitted over the channel. The receiver uses this sequence as its reference signal to adjust the coefficients of the adaptive filter to the channel. (After the training period, other methods are used to update the filter continually, which are beyond the scope of this exercise.)

(**a**) Test the performance of the channel compensator by generating the following signals:

(i) Use **f rgen** with seed value 3 to generate 1000 samples of a random sequence $r[n]$.

(ii) For the channel, use a filter with the system function

$$G(z) = \frac{1}{1 - 0.75z^{-1}}.$$

This filter can be generated with the *create file* option of **f siggen** and **f filter** can be used to generate 1000 samples of the channel output.

(iii) Let the length of the adaptive filter be $N = 9$.

Implement the adaptive filter using the function **f adaptfir** with an initial step size of $\mu = 0.1$ and initial coefficients that are zero. Record the final values of the coefficients in the adaptive filter.

(**b**) Repeat the operations in (a) but gradually reduce the length of the reference signal, M. Vary the length of the training sequence, M, and determine the smallest value of M that will allow the filter coefficients to converge to within 1% of the values that you found in (a).

(**c**) Vary μ to determine what effect it has on the convergence of the filter. What happens if μ is made too large? What value of μ maximizes the rate of convergence?   □

EXERCISE 2.6.6. Partial Noise Removal

Adaptive filters have been used for active noise control. Assume that signals are recorded from two microphones. The first records $r[n]$, which is our noise-corrupted signal. For example, $r[n]$ might be a speech signal in a noisy background, as might occur in a crowded room or in an aircraft cockpit:

$$r[n] = s[n] + \epsilon_1[n],$$

where $s[n]$ is the undisturbed speech and $\epsilon_1[n]$ is the background noise. The second microphone records only a corrupted version of the background noise:

$$x[n] = 0.9\epsilon_1[n] + \epsilon_2[n],$$

where $\epsilon_2[n]$ accounts for the measurement error in estimating $\epsilon_1[n]$. An adaptive filter is used to obtain $y[n]$, which is an estimate of the noise $\epsilon_1[n]$. The corrupted input signal $r[n]$ is then subtracted from the noise estimate to obtain $e[n] \approx -s[n]$ (the uncorrupted input). The complete system is shown in the diagram below.

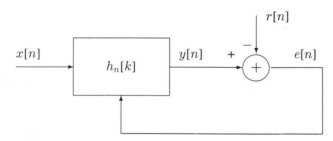

(a) In this part of the problem, you will need to generate several signals.

   (i) Use the *triangular wave* option of **f siggen** to generate 100 periods of a triangular wave with amplitude 1, period 10, and starting point $n = 0$. This is the signal $s[n]$.

   (ii) Use **f rgen** with a seed value of 4 and **f gain** to generate 1000 samples of a uniform random variable scaled by a factor of 0.1. Use **f filter** to convolve this signal with a filter with system function $H(z) = 1/(1 - 0.5z^{-1})$. Use the *create file* option of **f siggen** to create the filter and **f filter** to generate 1000 samples of the output. This is the signal $\epsilon_1[n]$.

   (iii) Use **f rgen** with a seed value of 5 and **f gain** to generate 1000 samples of a second uniform random sequence using a scale factor of 0.03. This is the signal $\epsilon_2[n]$.

   (iv) Use the function **f add** and **f gain** and the signals that you have already generated to generate the signals $r[n]$ and $x[n]$.

Measure and record the signal-to-noise ratio in the signal $r[n]$ using the function **f snr** by considering $s[n]$ to be the input signal.

(b) Implement the complete system shown in the figure. Let the adaptive filter length N be nine and use as initial values for the coefficients $h_0[k] = 0.0$ for $k = 0, 1, \ldots, 8$. Use $x[n]$ as the input to the filter and use $r[n]$ as the reference input. Use **f adaptfir** with $\mu = 0.03$ to implement the filter. Correct $e[n]$ by multiplying it by -1 using the **f gain** function. Display the last 100 samples of $e[n]$ and $s[n]$ on the same graph using **f extract** and **g view2** and comment on their appearance. Record the final values of the filter coefficients, $h_M[n]$.

(c) Measure the SNR for the signal $e[n]$. By how much has the signal been improved? You should observe a small improvement.   □

2.7 NONLINEAR FILTERING

Linear, time-invariant filters are not the only filters that are useful in applications. An example of a nonlinear FIR filter is the N-point median filter defined by the input–output relation

$$y[n] = \text{median}\{x[n - 2], x[n - 1], x[n], x[n + 1], x[n + 2]\},$$

where in this case $N = 5$. The median operation ranks its input values according to their amplitude and selects as its output the middle value. In this case, the output value would be the third largest of the five input samples centered around the current index. Median filters can be defined over larger or smaller data windows, although these windows are normally chosen to be of odd length. These are particularly effective at removing isolated noise spikes in the input sequence while preserving discontinuities.

Median filters are a special case of rank (or order statistics) filters. Other filters of this class replace the median operation by a maximum or minimum or a 'choose the Rth largest' operator. In the language of *mathematical morphology*, the minimum operation is a special case of an operation known as *erosion* and the maximum operation is a special case of *dilation*.

The function **f rank** will allow you to perform rank order filtering in a convenient way. Its filter parameters are N (the number of samples involved in the ordering), and R (the rank value). The function **f rank** orders the N samples in the window from lowest to highest. When $R = 1$, the function chooses the minimum value among the samples in the window, i.e., the first sample value in the rank ordered list. This case in which $R = 1$ is the "minimum" operation. When $R = N$, the function selects the largest value, or equivalently, the last sample value in the rank ordered list, which is the "maximum" operation.

EXERCISE 2.7.1. Rank Filtering

This exercise will make use of several 5-point filters. These are defined as

$$
\begin{aligned}
\text{mean:} \quad & y[n] = (x[n - 2] + x[n - 1] + x[n] + x[n + 1] + x[n + 2])/5 \\
\text{median:} \quad & y[n] = \text{median}\{x[n - 2], x[n - 1], x[n], x[n + 1], x[n + 2]\} \\
\text{maximum:} \quad & y[n] = \max\{x[n - 2], x[n - 1], x[n], x[n + 1], x[n + 2]\} \\
\text{minimum:} \quad & y[n] = \min\{x[n - 2], x[n - 1], x[n], x[n + 1], x[n + 2]\}
\end{aligned}
$$

They can be implemented using the functions **f convolve**, **f median**, and **f rank**. The output samples of **f rank** are formed by ranking the samples in a sliding N point window in order of increasing amplitude. The Rth largest is then chosen. When $R = N$, **f rank** performs a maximum operation; when $R = 1$, it performs a minimum operation; when $R = (N + 1)/2$, it selects the median.

(a) Create a signal using the functions **f siggen** and **f rgen** that is 64 points long consisting of a sum of three components—a square wave, an impulse train, and low-level white noise. Generate the square wave and the impulse train using **f siggen**. Design the square wave to consist of eight periods with a pulse length of four, a period of eight samples, and an amplitude of one. Let that signal begin at $n = 0$. Let the impulse train have an amplitude of 0.5, a period of 8, and a starting sample at $n = 0$. Use **f rgen** with seed value one to generate a random signal and then adjust its maximum amplitude to be 0.2 using **f gain**. Add the three components together using **f add** twice. For our purposes, we will consider the square wave to be a signal of interest and the other two signals to represent additive noise.

Filter this input sequence with each of the four filters above. For the mean use **f convolve** with the appropriately chosen impulse response, and for the other three you can use the function **f rank**. Display and sketch the results using **g view**. Describe the effects of the four filters on the input signal.

(b) In running the experiments above, you should have observed that the median filter removed most of the impulse train from the input. Find an operator (or combination of operators) whose output will preserve the impulse train but will remove the square wave. Implement your operation and display and sketch your result.   □

EXERCISE 2.7.2.   **Root Signals**

A *root signal* is a signal that is unchanged by a filtering operation. The all-zero sequence is trivially a root signal for all of the nonlinear operations that have been discussed in this section. This exercise will look for other root signals.

Begin by designing a composite signal **xn** composed of a triangle, a block, and a random component. Use **f siggen** to design a triangular signal consisting of one period of length 16 with amplitude 1 and starting point $n = 0$. Next, design a square wave with a pulse length equal to 16, a period length of 32, number of periods equal to one, amplitude 1, and starting point $n = 0$. For the last component of the signal, use **f rgen** to generate 32 samples of a random sequence with seed value 1. Use **f lshift** to shift the random signal to begin at $n = 48$. Add the random signal, the square wave, and the triangle together using **f add** to form **xn**. Display **xn** using **g view** and sketch the result.

Notice that there are four distinct regions represented in the signal: a triangular section where sample values increase and decrease monotonically, a flat constant region with unity amplitude, a zero value region, and a region of random values.

(a) Find the root signal of **xn** for the three-point median filtering operation and sketch the result. This can be done by applying the median filter repeatedly until the output converges. You can use **g view2** to check successive filtered outputs visually until differences are no longer present. How would you characterize the root signal for each of the four regions in **xn**? How do the root signal characteristics change with the filter length? Try finding the root signals using five-point and seven-point median filters before answering this question.

(b) Repeat part (a) for the maximum (dilation) operation, which can be implemented using **f rank**. This function is discussed in the introduction to this subsection.

(c) Repeat part (a) for the minimum (erosion) operation, which can also be implemented by using **f rank**. $\qquad\square$

REFERENCES

[1] J. F. Kaiser, "Nonrecursive Digital Filter Design Using the I_0-sinh Window Function," *Proc. 1974 Int. Symp. Circuits and Systems,* pp. 20–23.

[2] J. F. Kaiser and R. W. Hamming, "Sharpening the Response of a Symmetric Nonrecursive Filter by Multiple Use of the Same Filter," *IEEE Trans. Acoustics, Speech, Signal Processing,* Vol. 25 (Oct. 1977), pp. 415–422.

[3] J. W. Tukey, *Exploratory Data Analysis,* Addison-Wesley: Reading, MA, 1977.

[4] A. V. Oppenheim and R. W. Schafer, *Discrete-Time Signal Processing,* Prentice-Hall: Englewood Cliffs, NJ, 1989.

[5] T. W. Parks and J. H. McClellan, "Chebyshev Approximation for Nonrecursive Digital Filters with Linear Phase," *IEEE Trans. Circuit Theory,* Vol. 19 (Mar. 1972), pp. 189–194.

[6] O. Herrmann, L. Rabiner, and D. S. Chan, "Practical Design Rules for Optimum Finite Impulse Response Lowpass Digital Filters," *Bell Syst. Tech. Journal,* Vol. 52, No. 6. (July-August 1973), pp. 767–799.

[7] R. W. Hamming, *Digital Filters,* 3rd ed., Prentice-Hall: Englewood Cliffs, NJ, 1988.

[8] B. Widrow and S. D. Stearns, *Adaptive Signal Processing,* Prentice-Hall: Englewood Cliffs, NJ, 1985.

[9] M. L. Honig and D. G. Messerschmidt, *Adaptive Filters: Structures, Algorithms, and Applications,* Kluwer Academic Publishers: Hingham, MA, 1984.

[10] S. Haykin, *Introduction to Adaptive Filters,* Macmillan: New York, 1984.

[11] S. Haykin, *Adaptive Filter Theory,* Prentice-Hall: Englewood Cliffs, NJ, 1986.

[12] B. Widrow et al., "Stationary and Nonstationary Learning Characteristics of the LMS Adaptive Filter," *Proc. IEEE,* Vol. 64 (Aug. 1976), pp. 1151–1161.

IIR Filters 3

IIR (Infinite length Impulse Response) filters are important components in a variety of systems for discrete-time processing. They offer a number of advantages over FIR filters in certain applications because of their excellent magnitude response characteristics, particularly when the width of a passband or stopband must be very narrow or when very narrow transition bands or high attenuations are required.

IIR filters can be implemented efficiently using sets of recursive difference equations of the form

$$y[n] = - a_1 y[n - 1] - a_2 y[n - 2] - \cdots - a_N y[n - N]$$
$$+ b_0 x[n] + b_1 x[n - 1] + \cdots + b_M x[n - M] \qquad (3.1)$$

in which $x[n]$ is the input sequence and $y[n]$ is the output sequence. When N consecutive samples of $y[n]$ (known as *initial conditions*) are known, (3.1) can be used to calculate all of the successive output samples. This type of realization, in which previously computed output samples are used to compute future ones, is called *recursive*. The terms *IIR filter* and *recursive filter* are often used synonymously because (3.1) can realize both, but technically, they are different. The adjective "IIR" refers to the form of the impulse response of the filter while the adjective "recursive" tells how it is implemented. There are recursive implementations for FIR filters, and there are IIR filters that cannot be implemented recursively. Nonetheless, we will use these terms interchangeably in this chapter without ambiguity, since we will always be discussing IIR filters described by difference equations of the form given in (3.1).

Recursive filters are conveniently defined by their system functions. (Recall that the system function is the z-transform of the impulse response of the filter.) These are rational functions in the variable z^{-1}. The system specified by (3.1) has the system function

$$H(z) = \frac{\sum\limits_{m=0}^{M} b_m z^{-m}}{1 + \sum\limits_{n=1}^{N} a_n z^{-n}} \tag{3.2}$$

if it is at initial rest (i.e., if it has zero initial conditions). This can also be written in factored form as

$$H(z) = \gamma_0 \frac{\prod\limits_{m=1}^{M}(1 - \beta_m z^{-1})}{\prod\limits_{n=1}^{N}(1 - \alpha_n z^{-1})}. \tag{3.3}$$

The roots of the numerator polynomial, β_m, are called the *zeros* of the filter, and the roots of the denominator polynomial, α_n, are called the *poles*. The final parameter γ_0 is a constant gain. In general, the number of poles and zeros is directly related to the effort required to implement the filter; it is also related to how well the frequency response of the filter can approximate an ideal response. The *filter order*, N, for an IIR filter is the number of poles that it contains in the finite z-plane.

A linear, time-invariant filter is *causal* if its impulse response is zero for $n < 0$. The current output sample of a causal filter depends only on the current and previous samples of the input. When we limit the inputs to the filter to those for which $x[n] = 0$, for $n < 0$ and initialize (3.1) with the initial conditions $y[-1] = y[-2] = \ldots = y[-N] = 0$, the recursive implementation in (3.1) results in a causal filter. Causality is an important constraint for filters that must operate in a real-time environment, where at each tick of a clock one input sample arrives and one output sample must be produced. In certain non-real-time applications, where the entire input signal has been received before any processing begins, noncausal filters can be used.

An even more important constraint on an IIR filter is that it be *stable*. If a filter is unstable, its output sequence can grow without bounds even when the input signal is well behaved. Whether or not a filter is stable depends on the locations of its poles; a causal LTI IIR filter is stable if and only if each pole satisfies the condition $|\alpha_n| < 1$. This means that all of its poles must lie inside the unit circle in the z-plane. This chapter will focus attention on the design of stable, causal IIR filters. We should note that all FIR filters are stable. This is one advantage that FIR filters have over IIR filters.

One of the most popular approaches to recursive digital filter design is to first design an analog filter, and then transform it to a digital one. This approach is attractive for two reasons: first because the analog filter design problem is so well understood, and second because mappings exist that can map certain optimal analog filters into optimal digital ones. Since classical analog filter design techniques may be unfamiliar, our treatment of the design problem begins with a summary of these methods followed by a discussion of methods for transforming these filters into digital ones. There are other design approaches, which we will not discuss, where filters are designed directly in the discrete-time domain. These methods involve numerical optimization and are usually computationally intensive. A discussion of some of these methods can be found in [1],

[2], and [3].

When designing either IIR or FIR filters, we usually adhere to performance specifications related to the maximum passband and stopband magnitude error, the passband and stopband cutoff frequencies, and the filter order. For classical analog filter design, some of these terms are defined differently. In this text, we will usually assume that the passband response of a digital filter lies within δ_p of its nominal value, i.e., that the response varies between $1 + \delta_p$ and $1 - \delta_p$ when the nominal gain of the filter is one. This is the common convention for digital filters. For analog filter designs, the popular convention is to assume that the maximum gain is 1 and that the passband magnitude response varies between 1 and $1 - \Delta_p$. It is not difficult to convert from one convention to the other, but we do need to be aware of the difference.

3.1   ANALOG FILTER DESIGN

This section will look at classical analog Butterworth, Bessel, Chebyshev, and elliptic filters. All of these have system functions that are rational functions of the *Laplace transform* variable s,

$$H_a(s) = \frac{\sum_{k=0}^{M} c_k s^k}{s^N + \sum_{i=0}^{N-1} d_i s^i}. \tag{3.4}$$

The frequency response is equal to the system function with $s = j\Omega$ and has the form

$$H_a(j\Omega) = \frac{\sum_{k=0}^{M} c_k \, (j\Omega)^k}{(j\Omega)^N + \sum_{i=0}^{N-1} d_i \, (j\Omega)^i}. \tag{3.5}$$

The continuous-time frequency variable, Ω, has units of radians per second (rad/s). The design procedures that we shall study tell us how to determine the parameters M, N, c_k, and d_i from the filter specifications for a variety of filter types.

You will notice that in this chapter we are using the notation $H_a(j\Omega)$ to denote the analog frequency response rather than $H_a(\Omega)$. This is so that we can use the substitution $s \Leftrightarrow j\Omega$ to convert between the system function $H_a(s)$ and the frequency response, $H_a(j\Omega)$, without ambiguity.

The system function is the (bilateral) Laplace transform of the impulse response, $h_a(t)$, of the filter.[1] The Laplace transform of a continuous-time signal is analogous to the z-transform of a discrete-time sequence. It is defined by the integral

$$H_a(s) = \int_{-\infty}^{\infty} h_a(t) e^{-st} dt. \tag{3.6}$$

---

[1]The unilateral and bilateral Laplace transforms are identical if the system is causal.

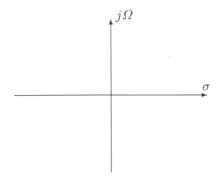

Figure 3.1. The s-plane.

Similarly, the frequency response is defined as

$$H_a(j\Omega) = \int\limits_{-\infty}^{\infty} h_a(t)e^{-j\Omega t}dt. \tag{3.7}$$

The frequency response can be inverted to determine the system impulse response by using the inverse Fourier transform integral

$$h_a(t) = \frac{1}{2\pi} \int\limits_{-\infty}^{\infty} H_a(j\Omega)e^{j\Omega t}d\Omega. \tag{3.8}$$

The Laplace transform variable, s, is a complex variable that is often expressed in terms of its real and imaginary parts σ and Ω:

$$s = \sigma + j\Omega.$$

Every value of s is equivalent to a point in a complex plane, known as the s-plane, which is shown in Figure 3.1. Since $H_a(j\Omega) = H_a(s)|_{s=j\Omega}$, it is common to say that the frequency response is equal to the system function evaluated on the imaginary (or frequency) axis. If the filter is causal and stable, then all of its poles must lie in the left-half plane (LHP) defined by $\sigma < 0$.

Analog design procedures determine the filter parameters that optimize the squared magnitude $|H_a(j\Omega)|^2$ of the frequency response. There are two reasons for this: first, the squared magnitude function is real, which is demanded by the mathematics of the optimizations being performed; second, the squared magnitude function is itself a rational function of s. We can show this by computing the inverse Fourier transform of $G_a(j\Omega) = |H_a(j\Omega)|^2$. Since

$$G_a(j\Omega) = |H_a(j\Omega)|^2 = H_a(j\Omega)H_a^*(j\Omega),$$

if $h_a(t)$ is real, its inverse Fourier transform is

$$g_a(t) = h_a(t) * h_a(-t).$$

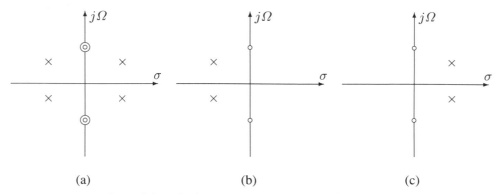

Figure 3.2. Spectral factorization. (a) Pole-zero plot for $G_a(s)$. (b) Pole-zero plot for $H_a(s)$. (c) Pole-zero plot for $H_a(-s)$.

By calculating the Laplace transform we see that $G_a(s) = H_a(s)H_a(-s)$, which is rational if $H_a(s)$ is rational. If $h_a(t)$ is real, $G_a(s)$ can be determined from the squared magnitude function directly by replacing $j\Omega$ by s. In general, $G_a(s)$ has poles and zeros that are distributed throughout the s-plane, but for every singularity in the right-half plane there will be one in the left-half plane and vice versa. (And every zero on the imaginary axis will be a double zero.)

The task of recovering $H_a(s)$ from $G_a(s)$ is known as *spectral factorization*. The process is straightforward, but the result is not unique. For every pole or zero in $G_a(s) = H_a(s)H_a(-s)$, we identify the pole or zero that is symmetrically located on the other side of the $j\Omega$-axis. One of these poles (or zeros) is assigned to $H_a(s)$ and the other to $H_a(-s)$. For $H_a(s)$ to correspond to a causal stable filter, the poles assigned to $H_a(s)$ should be those in the left-half plane; those in the right-half plane should be assigned to $H_a(-s)$. If the same rule is applied to the distribution of the zeros, as is commonly done, the resulting filter is called *minimum phase*. Double zeros on the imaginary axis should be split with one zero going to each of the system functions. This procedure is illustrated in Figure 3.2.

Before we begin the exercises, it is appropriate to comment on how analog system functions should be specified in signal files. We have been using signal files to represent discrete-time system functions, and it is important that our representations be consistent so that common routines can be used for such tasks as finding poles and zeros.

The analog filters used in the software are expressed in direct form and in terms of positive powers of s. They have the form

$$H_a(s) = \frac{c_P s^P + c_{P-1}s^{P-1} + \cdots + c_1 s + c_0}{s^Q + d_{Q-1}s^{Q-1} + \cdots + d_1 s + d_0}, \tag{3.9}$$

where the c_k's and d_k's are the analog filter coefficients. Note that $H_a(s)$, like $H(z)$, is a ratio of numerator and denominator polynomials. The two major differences are that the polynomials are in the variable s instead of z^{-1} and that the order of the coefficients is reversed. In the case of $H(z)$, the coefficients are in ascending order in z^{-1} while for

$$
\begin{array}{ll}
P + Q + 1 & \text{length} \\
P & \text{numerator order} \\
Q & \text{denominator order} \\
T & \text{1-real or 2-complex} \\
S & \text{starting point}
\end{array}
$$

NUMERATOR

c_P

$\vdots$

c_0

DENOMINATOR

d_{Q-1}

$\vdots$

d_0

Figure 3.3. The signal file corresponding to a continuous-time system function.

$H_a(s)$ they are in descending order in s. However, in terms of the DSP file structure, there is no difference. The general analog system function of (3.9) has a file structure as shown in Figure 3.3, which is the same as the discrete-time case. The numerator coefficients are entered in order, the leading one 1 in the denominator is not included, and the remaining denominator coefficients are also entered in order. Therefore, the same functions that were used to process discrete signal and filter files can be used to process the analog filter files.

EXERCISE 3.1.1.   **Spectral Factorization in the s-Domain**

The purpose of this exercise is to become more familiar with the process of spectral factorization. Let $G_a(j\Omega)$ be the squared magnitude function of the analog filter defined by

$$
G_a(j\Omega) = |H_a(j\Omega)|^2,
$$

where

$$
G_a(s) = \frac{1}{s^6 - 1} = \frac{1}{\displaystyle\prod_{\ell=0}^{5}(s - e^{j\pi\ell/3})}.
$$

We see that $G_a(s)$ has six poles that occur at $s = 0.5 \pm j0.866$, $s = -0.5 \pm j0.866$, $s = 1$, $s = -1$.

(a) Use the *change pole/zero* option in **g apolezero** to add the conjugate pole pairs $s = 0.5 \pm 0.866$ and $s = -0.5 \pm 0.866$, and the single poles $s = 1$ and $s = -1$. Sketch the magnitude response that appears on the screen.

(b) By inspection, determine and sketch the poles and zeros of $H_a(s)$ using the spectral factorization procedure described in this section. Since only poles are present in this filter, $H_a(s)$ contains only the poles in the left-half plane. Now use the *change pole/zero* option in **g apolezero** to delete the unwanted roots of $G_a(s)$ in the right-half plane. What remains after deletion is $H_a(s)$. Sketch $|H_a(j\Omega)|$ and notice that the magnitude of $G_a(j\Omega)$ is exactly the square of the magnitude of $H_a(j\Omega)$.   □

EXERCISE 3.1.2. Spectral Factorization in the z-Domain

The spectral factorization procedure that we developed for determining a system function from its squared magnitude function can be adapted for discrete-time system functions. This exercise develops the procedure and does not require the use of the computer.

Let the system function of the discrete-time system be $H(z)$, let its frequency response be $H(e^{j\omega})$, and let its impulse response be $h[n]$.

(a) Show analytically that the inverse DTFT of $G(e^{j\omega}) = |H(e^{j\omega})|^2$ is $g[n] = h[n] * h^*[-n]$ and that the corresponding z-transform is $G(z) = H(z)H^*(1/z^*)$, where the superscript * is used to denote the complex conjugate.

(b) If

$$H(z) = \frac{1 - 0.2z^{-1}}{(1 - 0.5jz^{-1})(1 + 0.5jz^{-1})},$$

sketch a pole/zero plot for $G(z) = H(z)H^*(1/z^*)$. Identify which poles are associated with $H(z)$ and which belong to $H^*(1/z^*)$.

(c) Determine $H(z)$ for the following squared magnitude function

$$G(e^{j\omega}) = \frac{1 + \cos \omega}{1 - 0.5 \cos \omega}.$$

□

EXERCISE 3.1.3. Minimum Phase FIR Filters

In Chapter 2, we considered a number of procedures for designing linear-phase FIR digital filters. A linear-phase characteristic is often desirable because it introduces no phase distortion, but has the disadvantage that the overall delay is usually long. If linear phase is not as important as the overall delay, a minimum-phase FIR filter might prove to be useful. These can be designed using the following procedure:

- Design an equiripple linear-phase FIR filter of odd length using **f pksmcc**. This will be called the *linear-phase prototype*.

- Add an impulse to the central sample of such a height that the overall frequency response is everywhere nonnegative. This can be done by using **g dfilspec** to determine the amplitude of the maximum ripple, and a text editor to add that value to the central sample. The modified linear-phase prototype will be called $G(e^{j\omega})$. An example is shown in Figure 3.4.

- Extract $H(z)$ from $G(z)$ by spectral factorization using **g polezero** and following the procedure introduced in Exercise 3.1.2. The spectral factorization should retain the zeros inside the unit circle and split the double zeros on the unit circle. It may be necessary to renormalize the gain.

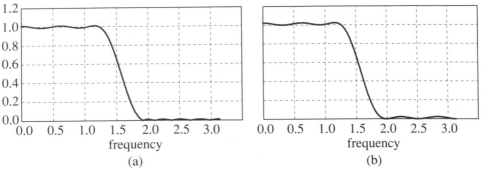

Figure 3.4. (a) A 21-point equiripple filter with impulse response $h[n]$ and stopband ripple $\delta_s = 0.0114$. (b) The frequency response of the filter $h[n] + 0.0114\delta[n - 10]$.

(a) Determine the length, cutoff frequencies, and ripples of the minimum-phase filter, $H(e^{j\omega})$, if the prototype filter designed by the Parks–McClellan algorithm has length $2N - 1$, passband and stopband cutoff frequencies ω_p and ω_s, respectively, and passband and stopband ripples δ_p and δ_s. Assume that the prototype filter has a nominal passband gain of 1 and that the final design is to have the same nominal gain.

(b) Use the appropriate software functions to design a minimum phase equiripple filter of minimum length to satisfy the following specifications:

$$\delta_p = 0.1,$$
$$\delta_s = 0.1,$$
$$\omega_p = 0.2\pi,$$
$$\omega_s = 0.4\pi.$$

Note that these specifications apply to the final minimum-phase filter, not to the linear-phase prototype. Record the actual passband and stopband ripples that you achieve.

(c) Design a *linear-phase* equiripple filter, $h[n]$, using **f pksmcc** to satisfy the same specifications. Again record the passband and stopband ripples. How do the lengths of the two filters compare?

(d) As an alternative, create a minimum-phase filter from the linear-phase filter, $h[n]$, that you designed in part (c). The minimum phase filter will have all its zeros either inside or on the unit circle. Create this filter by finding all of the zeros of $H(z)$ that do not lie on the unit circle and replacing them with zeros at their reciprocal locations. A convenient way to do this is to use **g polezero**. Recall that this function allows you to read in files, delete selected zeros, and output the modified results. Use **g polezero** to decompose $h[n]$ into two filters—one filter, $f[n]$, which contains the zeros of $h[n]$ that are located inside and on the unit circle, and another filter, $e[n]$, which contains the zeros outside the unit circle. Then use **f reverse** to create the sequence $e[-n]$, which will have its roots at the reciprocals of the roots of $e[n]$.

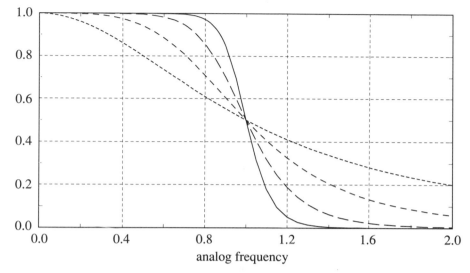

Figure 3.5. Butterworth squared magnitude responses for $N = 1, 2, 4, 8$.

The final step is to use **f convolve** to convolve $f[n]$ and $e[-n]$, and **f lshift** to make the result causal. Call the resulting filter $g[n]$. Is $|G(e^{j\omega})|$ equal to the magnitude of the filter that you found in part (b)? Which filter is better in terms of magnitude response characteristics?   □

3.1.1 Analog Butterworth Filters

The squared magnitude response of a Butterworth lowpass filter of order N is

$$|H_a(j\Omega)|^2 = \frac{1}{1 + (\frac{\Omega}{\Omega_c})^{2N}},$$ (3.10)

where Ω_c is the 3-dB cutoff frequency.[2] Replacing $j\Omega$ by s gives $G_a(s)$ where

$$G_a(s) = H_a(s)H_a(-s) = \frac{1}{1 + (\frac{s}{j\Omega_c})^{2N}}.$$ (3.11)

Figure 3.5 displays this squared magnitude response for several values of N. The filter $H_a(s)$ has the property that its magnitude response is maximally flat; the first $2N - 1$ derivatives of $|H_a(j\Omega)|^2$ are zero at $\Omega = 0$ and $\Omega = \infty$. Consequently, the magnitude response is monotonic and free of ripples in the passband and stopband. This figure also illustrates the fact that $|H_a(j\Omega)|$ more closely approximates the ideal filter as the order is increased. At the cutoff frequency Ω_c, notice that $|H_a(j\Omega_c)|^2 = 1/2$ for all N.

---

[2]The 3-dB cutoff is the frequency at which $|H_a(j\Omega)|$ is down by 3 decibels or, equivalently, equals $\sqrt{2}/2$. This cutoff frequency is also known as the half-power point because it is the frequency at which $G_a(j\Omega)$ (which is proportional to the power) equals $1/2$.

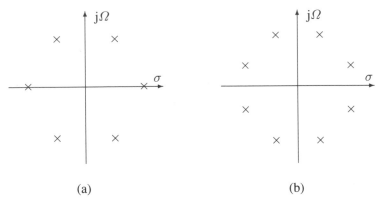

Figure 3.6. The poles of $H_a(s)H_a(-s)$ for a Butterworth filter. (a) Odd N. (b) Even N.

The desired filter performance is defined by its passband and stopband cutoff frequencies, its passband and stopband errors, and its order. These can be related to the two parameters Ω_c and N that defined the Butterworth filter in (3.10). If we require that the specification be satisfied exactly at the stopband edge, then

$$\Delta_s^2 = \frac{1}{1 + (\frac{\Omega_s}{\Omega_c})^{2N}}.$$ (3.12)

Solving for N yields

$$N = \frac{\log_{10}(\Delta_s^{-2} - 1)}{2\log_{10}(\frac{\Omega_s}{\Omega_c})}.$$ (3.13)

The poles of $H_a(s)H_a(-s)$ for a Butterworth filter lie on a circle of radius Ω_c as shown in Figure 3.6. The poles are equiangularly spaced with an angular spacing of π/N. For odd values of N, there are two poles on the real axis; for even values of N, the poles straddle the axes.

Spectral factorization extracts $H_a(s)$ by retaining the left-half plane poles. This gives the system function:

$$H_a(s) = \frac{C}{\displaystyle\prod_{\ell=1}^{N}(s - j\Omega_c e^{j(2\ell-1)\pi/2N})},$$ (3.14)

where C is a normalizing constant. This expression is very convenient for designing Butterworth filters. It is implemented in the function **f abutter** with Ω_c set to the value 1.[3]

---

[3]Normalizing the 3-dB cutoff frequency to unity in the design formula is a very common practice. We shall see later that it is a relatively simple matter to alter this cutoff frequency after the filter is designed, by using a simple frequency transformation.

3.1.2 Analog Bessel Filters

Bessel filters are all-pole filters with a number of characteristics that make them unique. They have a maximally flat group delay response and a step response that has a very small overshoot, typically less than one percent. The shapes of both the magnitude response and the impulse response approach a Gaussian as N becomes large. These are seen in the Bessel filter plots in Figure 3.7.

The system function of an Nth-order Bessel filter is given by

$$H_a(s) = \frac{(2N)!}{2^N N! B_N(s)}, \tag{3.15}$$

where $B_N(s)$ is the Nth-order Bessel polynomial defined by

$$B_N(s) = \sum_{\ell=0}^{N} \frac{(2N - \ell)!}{2^{N-\ell}\ell!(N - \ell)!} s^\ell. \tag{3.16}$$

The Bessel polynomials can be evaluated conveniently using the recurrence formula

$$B_N(s) = (2N - 1)B_{N-1}(s) + s^2 B_{N-2}(s),$$
$$B_0(s) = 1,$$
$$B_1(s) = s + 1.$$

The Bessel filter is somewhat unusual because its 3-dB point varies as a function of the filter order, which is the only free parameter. This is less of a problem than it might appear, however, because the cutoff frequency can be modified by using frequency scaling (which is introduced in Exercise 3.1.6 and discussed further in the next section).

3.1.3 Analog Chebyshev I Filters

Chebyshev Type I filters are all-pole filters that have an equiripple magnitude response in the passband and a monotonic roll-off in the stopband as shown in Figure 3.8a. These filters are optimal in the sense that there is no other Nth-order all-pole filter with the same transition band that performs better in the passband and stopband.

The squared magnitude response of a Chebyshev Type I filter is given by

$$|H_a(j\Omega)|^2 = \frac{1}{1 + \epsilon^2 T_N^2(\Omega/\Omega_p)}, \tag{3.17}$$

where $T_N(x)$ is the Chebyshev polynomial of degree N. The magnitude response, $|H_a(j\Omega)|$, in the passband oscillates between a maximum value of 1 and a minimum value of $1/\sqrt{1 + \epsilon^2}$. The parameter Ω_p is the passband cutoff frequency. The Chebyshev polynomials themselves oscillate between -1 and 1 for values of x between -1 and 1. They are formally defined by the relation

$$T_N(x) = \begin{cases} \cos(N \cos^{-1} x) & |x| \le 1 \\ \cosh(N \cosh^{-1} x) & |x| > 1, \end{cases} \tag{3.18}$$

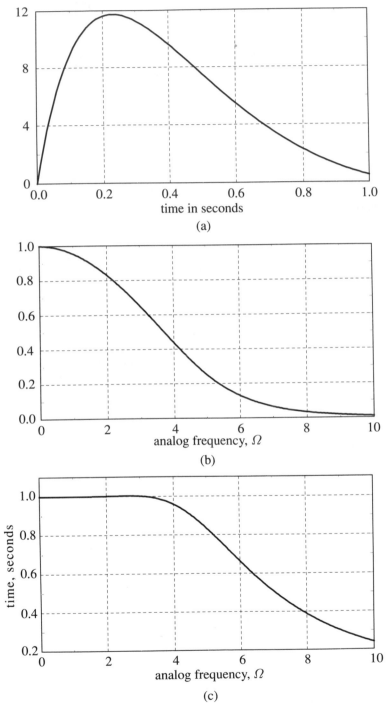

Figure 3.7. A sixth-order Bessel filter. (a) Impulse response. (b) Magnitude response. (c) Group delay response.

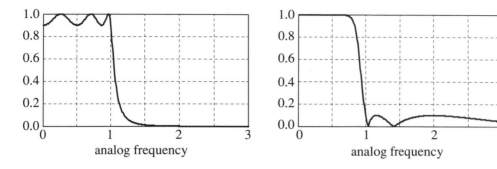

Figure 3.8. Magnitude response of analog Chebyshev filters. (a) Type I. (b) Type II.

but are more easily generated by using the recursion

$$T_{N+1}(x) = 2xT_N(x) - T_{N-1}(x) \tag{3.19}$$
$$T_0(x) = 1$$
$$T_1(x) = x.$$

Equation (3.17) can be easily expressed in terms of the Laplace transform variable s as

$$G_a(s) = H_a(s)H_a(-s) = \frac{1}{1 + \epsilon^2 T_N^2(s/j\Omega_p)}. \tag{3.20}$$

$H_a(s)$ is extracted from $G_a(s)$ by spectral factorization.

The poles of the Chebyshev filter lie on an ellipse in the s-plane and are uniformly spaced in angle. This leads to an alternative design formula that specifies the pole locations. It is given by

$$H_a(s) = \frac{D}{\displaystyle\prod_{\ell=1}^{N}(s - s_\ell)}, \tag{3.21}$$

where

$$s_\ell = \Omega_p\frac{(\gamma^{-1} - \gamma)}{2} \sin\frac{(2\ell - 1)\pi}{2N} + j\Omega_p\frac{(\gamma^{-1} + \gamma)}{2} \cos\frac{(2\ell - 1)\pi}{2N} \tag{3.22}$$

$$\gamma = \left(\frac{1 + \sqrt{1 + \epsilon^2}}{\epsilon}\right)^{1/N}, \tag{3.23}$$

and D is the constant of normalization. The system function of a Chebyshev Type I filter can be obtained easily by evaluating this expression, which removes the need for spectral factorization. This is done in the function **f acheby1**, which accepts as input parameters, N, the order, and ϵ, the passband ripple parameter. The passband ripple, Δ_p, is related to ϵ by

$$\Delta_p = 1 - \frac{1}{\sqrt{1 + \epsilon^2}}.$$

For simplicity of formulation, the passband cutoff frequency Ω_p is constrained to be one.

3.1.4 Analog Chebyshev II Filters

Chebyshev Type II (or Chebyshev II) filters are the complements of Chebyshev Type I filters. They have magnitude responses that are monotonic in the passband and equiripple in the stopband, as illustrated in Figure 3.8b. Unlike the Type I filters, which are all-pole, the Type II filters contain both poles and zeros. They are described by the squared magnitude response

$$|H_a(j\Omega)|^2 = \frac{\epsilon^2 T_N^2(\Omega_s/\Omega)}{1 + \epsilon^2 T_N^2(\Omega_s/\Omega)}. \tag{3.24}$$

The parameter, ϵ, defines the maximum stopband deviation; Ω_s is the stopband cutoff frequency. The Type II filters, like the Type I filters, use the Chebyshev polynomials. These can be computed using either (3.18) or (3.20). Alternatively, the system function can be specified directly from the zero and pole locations. This approach has the advantage that it does not require polynomial factorization.

All of the zeros lie on the imaginary axis at the values

$$r_k = j\frac{\Omega_s}{\cos([2k-1]/2N)\pi}, \quad k = 1, 2, \ldots, N.$$

When N is odd, one of the zeros lies at infinity. The poles can be shown [2] to lie at $s_k = \sigma_k + j\Omega_k$, $k = 1, 2, \ldots, N$, where

$$\sigma_k = \frac{\Omega_s \alpha_k}{\alpha_k^2 + \beta_k^2},$$

$$\Omega_k = \frac{\Omega_s \beta_k}{\alpha_k^2 + \beta_k^2},$$

and

$$\alpha_k = -\sinh\phi \sin\left[\frac{(2k-1)\pi}{2N}\right],$$

$$\beta_k = -\cosh\phi \cos\left[\frac{(2k-1)\pi}{2N}\right],$$

$$\sinh\phi = \frac{\gamma - \gamma^{-1}}{2},$$

$$\cosh\phi = \frac{\gamma + \gamma^{-1}}{2},$$

$$\gamma = \left(\frac{(1+\epsilon^2)^{1/4}}{\epsilon^{1/2}} + (\frac{(1+\epsilon^2)^{1/2}}{\epsilon} - 1)^{1/2}\right)^{1/N},$$

and

$$H_a(s) = B\frac{\displaystyle\prod_{k=1}^{N}(s - r_k)}{\displaystyle\prod_{k=1}^{N}(s - s_k)}.$$

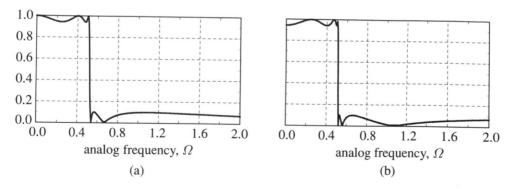

Figure 3.9. Two elliptic lowpass filters. (a) $N = 5$. (b) $N = 6$.

Thus the filter can be designed by specifying the parameters Ω_s, ϵ, and N. The ripple parameter ϵ is related to the maximum stopband deviation Δ_s by the expression

$$\Delta_s = \frac{\epsilon}{\sqrt{1 + \epsilon^2}}.$$

The fact that the Chebyshev II filters possess zeros represents both an advantage and a disadvantage when they are compared with Chebyshev I filters. Because of their zeros they are more complex to implement, but they have a more constant group delay. As we will see in a later section, the Chebyshev I and Chebyshev II filters that satisfy a given set of specifications will be of the same order.

3.1.5 Analog Elliptic Filters

Elliptic filters (sometimes called *Cauer filters*) are optimal in the same sense that linear-phase FIR filters designed by the Parks–McClellan algorithm are optimal. For a given set of passband and stopband cutoff frequencies no other filter of the same order has smaller passband and stopband errors. Also, like the Parks–McClellan filters, these are equiripple. An Nth-order elliptic filter will have N poles and it will usually have N zeros that are located on the imaginary axis of the s-plane.

Figure 3.9 displays elliptic magnitude responses for even and odd order lowpass filters. Notice that the response is equiripple in both the passband and the stopband—a property that is characteristic of elliptic filters. The squared magnitude response is given by

$$|H_a(j\Omega)|^2 = \frac{1}{1 + \epsilon^2 U_N^2(\Omega, L)}, \tag{3.25}$$

where ϵ is the passband deviation parameter and $U_N(\Omega, L)$ is the Nth-order Jacobian elliptic function. The parameter, L, contains information about the relative heights of the ripples in the passband and stopband. A proper treatment of elliptic functions is beyond the scope of this text. The interested reader is referred to [3] and [4]. For our purposes, it is sufficient to just accept that we have a program **f aelliptic** that evaluates the design equations for analog elliptic filters. The analog design parameters are Δ_p, Δ_s, and Ω_s. (Ω_p is constrained to be unity in the program.) The filter order is estimated from another set of equations and must also be specified. The function **f aelliptic** will suggest an appropriate order for you to use. Alternatively, **f oelliptic** can be used to estimate the filter order. The nature of the equations used in **f aelliptic** does not allow all specifications to be met precisely all the time. In some cases, when using the estimated filter order, the actual passband and stopband ripples may exceed the specifications. In other cases, the order estimation equations may underestimate the required order. Thus it is advisable to check the filter characteristics after the filter is designed to verify that the specifications are satisfied.

3.1.6   Estimating Filter Order

Filters can be specified in different ways. Sometimes the order, passband and stopband deviations, and center cutoff frequencies are specified, in which case the goal of the design procedure is to minimize the width of the transition band. On other occasions, the transition band cutoffs and passband/stopband deviations are specified and the goal is a filter of minimum order. The latter is often more difficult, however, because many of the compact analog design formulas require that the order be known. In this section, we summarize the formulas that specify the filter orders for the classical analog Butterworth, Chebyshev, and elliptic filters given a set of cutoff frequencies and frequency band deviations.

These formulas are normally stated in terms of two auxiliary parameters: η and k. The latter is called the *transition ratio*. For a lowpass filter it is defined simply as

$$k = \frac{\Omega_p}{\Omega_s}. \tag{3.26}$$

The parameter η has no formal name, but it is a basic analog filter parameter that is related to the ripple heights. It is defined by

$$\eta = \frac{\Delta_s 2\sqrt{\Delta_p}}{(1 - \Delta_p)\sqrt{(1 + \Delta_p)^2 - \Delta_s^2}}. \tag{3.27}$$

For a lowpass Butterworth filter, the required order is given by

$$N = \frac{\ln \eta}{\ln k}. \tag{3.28}$$

For a Chebyshev Type I or Type II lowpass filter, it is given by

$$N = \frac{\cosh^{-1}(1/\eta)}{\ln\left(\dfrac{1 + \sqrt{1 - k^2}}{k}\right)}. \tag{3.29}$$

Finally, for an elliptic filter the order is given by

$$N = \frac{K(k)K(\sqrt{1 - \eta^2})}{K(\eta)K(\sqrt{1 - k^2})}, \tag{3.30}$$

where $K(\cdot)$ is the complete elliptic integral of the first kind.

The Chebyshev and elliptic order estimation formulas are implemented by the functions **f ocheby1**, **f ocheby2**, and **f oelliptic**. A modified version of (3.26)–(3.28) with inputs expressed in terms of Ω_s, Δ_s, and a unity 3-dB cutoff frequency is implemented in the function **f obutter**. In particular, that function is based on (3.13).

EXERCISE 3.1.4.   **Analog Butterworth Filter Design**

In this exercise, you will design several Butterworth filters and examine their properties. Equation (3.11) provides a simple formula for $H(s)H(-s)$. The causal, stable analog Butterworth filter may be obtained by factoring this expression, discarding the roots in the right-half plane, and multiplying the remaining roots back together. Alternatively, (3.14) can be used to design the filter directly.

(a) Using (3.14) and without the use of the computer, design a second-order analog Butterworth filter $F_a(s)$ with $\Omega_c = 1$. Record the coefficients of $F_a(s)$ and show all of your work. Next, use **f abutter** to design the same filter. The function **f abutter** is a computer implementation of (3.14). Type the coefficients to the screen and verify that the hand-generated and computer-generated filters are the same. Notice that the first coefficient, a_0, in the denominator of the DSP filter file is always assumed to be 1 and therefore is not included in the DSP file, as illustrated in Figure 3.3.

(b) Using **f abutter**, design a tenth-order Butterworth lowpass filter, $H_a(s)$. Use **g apolezero** to display the poles of $H_a(s)$ and sketch them. Make another sketch corresponding to the poles of $H_a(s)H_a(-s)$.

(c) Display and sketch the magnitude response of $H_a(s)$ using **g afreqres** with $\Omega_L = 0$ and $\Omega_H = 2$. The function **g afreqres** produces a file called _fr_, which contains the actual complex signal that is used in the frequency plot. Use **f mag** to take the magnitude of _fr_ and write it to the file **temp1**. You may examine **temp1** by using **g look**. Do this and verify that it is a plot of the magnitude response of $H_a(j\Omega)$. As we showed earlier, the frequency response associated with $H_a(s)H_a(-s)$ is the squared magnitude of $H_a(j\Omega)$. Explicitly compute the squared magnitude $|H_a(j\Omega)|^2$ by typing

$$\textbf{f multiply} \quad \textbf{temp1} \quad \textbf{temp1} \quad \textbf{temp2}$$

which will store $|H_a(j\Omega)|^2$ in the file **temp2**. Now display and sketch $|H_a(j\Omega)|$ and $|H_a(j\Omega)|^2$ together by typing

$$\textbf{g look2} \quad \textbf{temp1} \quad \textbf{temp2}$$

Carefully sketch these plots from the screen and notice that the value of $|H_a(j\Omega)|$ at its 3-dB cutoff frequency (which appears at 90 on the **g look2** function axis) is 0.707, which corresponds to a 3-dB decrease in amplitude on the $20\log_{10}|H_a(j\Omega)|$ scale. Also observe that the value of $|H_a(j\Omega)|^2$ at this frequency is $1/2$. Since $|H_a(j\Omega)|^2$ can be viewed as the power of the signal, Ω_c is sometimes called the half-power point frequency.

(d) The normalizing constant C in (3.14) is a function of Ω_c in general. Analytically determine the value of C when $\Omega_c = 1$. Determine it when Ω_c is an arbitrary positive real number. □

EXERCISE 3.1.5.   **Cutoff Frequency of a Butterworth Filter**

One property of the Butterworth filter $H_a(s)$ is that the poles $H_a(s)H_a(-s)$ lie on a circle in the s-plane. This implies that the zeros of $H_a(s)$ lie on a half circle in the left half of the plane. What is the relationship between the radius of the circle and the cutoff frequency? Using this property, design a fourth-order Butterworth filter with $\Omega_c = 2$ by manually placing the poles of the filter in the s-plane. The *change pole/zero* option in **g apolezero** will allow you to create this filter by adding poles (of your choosing) to the s-plane. Use the *write to output file* option in the main menu to store your filter in a file. Display and sketch the magnitude and phase of this filter using **g afreqres** with $\Omega_L = 0$ and $\Omega_H = 8$.   □

EXERCISE 3.1.6.   **Frequency Scaling**

The function **f abutter** designs Butterworth lowpass filters with a 3-dB cutoff frequency of $\Omega_c = 1$. However, this function can be used to accommodate a wide range of cutoff frequencies by using a technique called *analog frequency scaling*.

(a) If Ω is replaced by Ω/a in (3.10), is the resulting filter still a Butterworth filter? How is the 3-dB cutoff frequency of the new filter related to the cutoff frequency of the old one? How are the filter coefficients related to the original ones?

(b) Consider the design of a fourth-order analog lowpass Butterworth filter, $H_a(s)$, with cutoff frequency $\Omega_c = 2\pi(10) = 62.832$ using the direct design formula given in (3.14). If we explicitly compute the real and imaginary parts of the four poles of the filter, i.e., $j\Omega_c e^{j(2\ell-1)\pi/8}$, where $\ell = 1, 2, 3, 4$, we obtain the two conjugate pole pairs $s = -24 \pm j58$ and $s = -58 \pm j24$. Use the *change pole/zero* option in **g apolezero** to add in these conjugate pole pairs to effectively create the filter. Return to the main menu and use the *write to output file* option to write the new filter to a file. Use **g afreqres** to display the log magnitude and magnitude responses in the range $\Omega_L = 0$ and $\Omega_H = 200$ (rad/s). Sketch these plots.

(c) Now design a fourth-order filter using **f abutter** and scale the coefficients appropriately using **f atransform**. The function **f atransform** allows you to change the s variable in the following way:

$$s \rightarrow \frac{c_0 + c_1 s + c_2 s^2}{d_0 + d_1 s + d_2 s^2}.$$

Only d_0 is of interest here and should be set equal to $2\pi(10) = 62.83$. Thus, specify $c_0 = 0$, $c_1 = 1$, $c_2 = 0$, $d_0 = 62.83$, $d_1 = 0$, and $d_2 = 0$. With these parameters, the transformation reduces to

$$s \rightarrow \frac{s}{2\pi(10)}$$

as desired. Compare the magnitude response of the resulting filter with that of the one that you designed in part (b) using **g afreqres** with $\Omega_L = 0$ and $\Omega_H = 200$ rad/s. Provide a sketch of the new plots. Comment on any differences that you observe and explain them.   □

EXERCISE 3.1.7. Analog Butterworth Filter Design

Consider the design of an analog Butterworth lowpass filter, $H_a(j\Omega)$, with magnitude deviation less than one percent in the region $|\Omega| < 8$ and $|H_a(j\Omega)|$ less than 0.01 in the region $|\Omega| > 14$.

(a) Determine the order of $H_a(j\Omega)$ using **f obutter**. (First you will need to determine a value for Ω_c. Then, using this value, calculate normalized passband and stopband cutoff frequencies.)

(b) Use **f abutter** and the concept of frequency scaling discussed in Exercise 3.1.6 to design the filter. The function **f atransform** can be used to perform the frequency scaling. Display and sketch its magnitude response using **g afilspec** with $\Omega_L = 0$ and $\Omega_H = 30$. Use the *right arrow* and *left arrow* keys to check the maximum passband and stopband deviations and stopband cutoff frequency.   □

EXERCISE 3.1.8. Determining the Butterworth Filter Order

There is one difficulty in using (3.13) to estimate the order of the Butterworth filter. This equation assumes that Ω_c is known. While this may sometimes be the case, it is more common for the filter specifications to be given in terms of Ω_p and Ω_s. In this exercise, you derive an alternative formula for estimating the filter order in terms of these parameters. The exercise does not require the use of a computer, although a calculator is helpful.

(a) Since the values of the squared magnitude function at $\Omega = \Omega_s$ and $\Omega = \Omega_p$ are known, we can write

$$\Delta_s^2 = \frac{1}{1 + (\Omega_s/\Omega_c)^{2N}} \tag{3.31}$$

and

$$(1 - \Delta_p)^2 = \frac{1}{1 + (\Omega_p/\Omega_c)^{2N}}. \tag{3.32}$$

Show that these two equations are equivalent to

$$\left(\frac{\Omega_p}{\Omega_s}\right)^{2N} = \frac{\Delta_s^2(1 - (1 - \Delta_p)^2)}{(1 - \Delta_s^2)(1 - \Delta_p)^2}. \tag{3.33}$$

(b) Apply (3.33) to determine the order of the lowest-order Butterworth filter that will satisfy the following specifications:

$$\Delta_p = 0.01,$$
$$\Delta_s = 0.001,$$
$$\Omega_p = 1.0,$$
$$\Omega_s = 2.0.$$

Remember that the order must be an integer.

(c) Once the order is known, either (3.31) or (3.32) can be used to determine the value of Ω_c. Determine the value of Ω_c using (3.31) and compare it with the value you get using (3.32). Why are these values different? Can you find a third value for Ω_c that will also satisfy the filter specifications? □

EXERCISE 3.1.9.   Improving on the Butterworth Filter Performance

The formula for the squared magnitude function for an analog Butterworth filter that was given in (3.10) has a maximum passband gain of 1. Since the magnitude response is monotonic, the passband gain will vary from 1 at $\Omega = 0$ to $1 - \Delta_p$ at $\Omega = \Omega_p$. Often, the filter specification will require merely that the passband response lie within Δ_p of unity. This means that in the passband region, the filter magnitude response is permitted to deviate both above and below unity as long as the deviation is not greater than Δ_p. The filter specified by (3.10) satisfies this condition, but it is overly conservative. We might be able to satisfy the specification with a lower-order filter by scaling the filter so that its peak gain is $1 + \Delta_p$ rather than 1. This exercise explores this possibility.

(a) Modify (3.33) for the case where the peak gain of the Butterworth filter is $1 + \Delta_p$ instead of 1. In your formula, you may ignore the effect of the scaling operation on the height of the stopband deviation, since that is a second-order effect.

(b) Using the formula that you developed in part (a), determine the lowest order Butterworth filter, $H_a(s)$, that will meet the following filter specifications.

$$\text{passband deviation} \qquad 1 - \Delta_p \leq |H_a(j\Omega)| \leq 1 + \Delta_p$$
$$\text{stopband deviation} \qquad |H_a(j\Omega)| \leq \Delta_s$$
$$\Delta_p = 0.01$$
$$\Delta_s = 0.01$$
$$\text{passband cutoff frequency, } \Omega_p = 1 \text{ rad/s}$$
$$\text{stopband cutoff frequency, } \Omega_s = 2 \text{ rad/s}$$

Use **f abutter**, **f atransform**, and **f gain** to design $H_a(s)$. Examine it carefully using **g afilspec** and verify that it satisfies the design specifications.

(c) Repeat part (b) when the maximum filter gain is restricted to 1 and compare the results. Summarize your observations. □

EXERCISE 3.1.10.   Bessel Filters

The function **f abessel** computes the Nth-order analog Bessel filter based on (3.15) and (3.16).

(a) Use **f abessel** to determine system functions of the analog Bessel filters of orders 2, 6, 10, and 14. Display and sketch the magnitude response of each filter using **g afilspec** with $\Omega_L = 0$ and $\Omega_H = 10$. Remember, the 3-dB cutoff frequency for a filter with a unity passband gain is the frequency at which the magnitude has a value of 0.707. Determine and record the 3-dB frequencies for each plot.

(b) Plot the data found in part (a). Find an approximate functional relationship between the 3-dB cutoff frequency and the order N. □

EXERCISE 3.1.11.   **Comparison of Butterworth and Bessel Filters**

Do Exercise 3.1.6, *Frequency Scaling*, if you have not already done so.

(a) Design an eighth-order Bessel filter using **f abessel**. Display and sketch the magnitude response of the filter using **g afilspec** with $\Omega_L = 0$ and $\Omega_H = 10$. Determine its 3-dB cutoff frequency and stopband cutoff frequency, Ω_s, defined as the frequency at which $H_a(\Omega) = 0.1$. Use the *frequency scaling* method discussed in Exercise 3.1.6 to move the 3-dB cutoff frequency to $\Omega_c = 1$. Use **g afilspec** to verify that your frequency scaling worked correctly.

(b) Use **g apolezero** to display the pole/zero plot of the frequency normalized analog Bessel filter and sketch it in the *s*-plane.

(c) Using **f obutter**, what order Butterworth filter would yield a filter with a comparable magnitude response characteristic? Design the filter using **f abutter** and sketch its group delay response from the **g afreqres** plot with $\Omega_L = 0$ and $\Omega_H = 10$. How does its group delay compare to that of the Bessel filter?   □

EXERCISE 3.1.12.   **Analog Chebyshev I Filters**

(a) The function **f acheby1** implements (3.21–3.23). As an introduction to using this function, design a seventh-order analog Chebyshev I lowpass filter with $\Omega_p = 1.0$ and $\epsilon = 0.3$.[4] Display and sketch the magnitude response and pole/zero plot using **g apolezero**. Use the *list pole/zero* option in **g apolezero** to list the poles and record these values.

(b) Use **g afilspec** to display the magnitude response with $\Omega_L = 0$ and $\Omega_H = 3$. How many ripples does the filter possess in the passband? Determine the frequencies at which the ripple peaks occur. Do the imaginary parts of the poles correspond to the frequencies of the passband ripples?

(c) Assume that the stopband is defined as those frequencies for which the magnitude response is less than 0.01. Use **g afilspec** to determine the stopband cutoff frequency, Ω_s, for the filter.   □

EXERCISE 3.1.13.   **Evaluating Stopband Performance**

The design parameters for an analog Chebyshev I filter are ϵ, N, and Ω_p. These are different from those that we have become used to. The purpose of this problem is to relate these parameters to the more familiar digital filter design specifications.

(a) The analog passband ripple, Δ_p, is related to ϵ in a simple way as shown in Figure 3.8. However, determination of the maximum stopband deviation, Δ_s, is more involved. Because the stopband behavior of the filter is monotonic, the values of Δ_s and Ω_s are tightly coupled and both depend upon the order N. Using the properties of the Chebyshev polynomials that we have discussed, derive an approximate formula that relates Δ_s, Ω_s, and N for fixed values of the other parameters. Test your formula by

---

[4]Remember that the function always designs filters with $\Omega_p = 1$. Thus you are not prompted for this value.

considering the design of a lowpass Chebyshev I filter with $\Delta_p = 0.005$, $\Delta_s = 0.05$, $\Omega_p = 1$, and $\Omega_s = 1.8$. Determine the order analytically and show your work. Verify your answer by using **f ocheby1**.

(b) The convention for digital filters is that the ripples oscillate above and below unity by an amount, δ_p, in the passband. However, for classical *analog* filters, the maximum passband value is typically fixed at unity and the passband deviation occurs between 1 and $1 - \Delta_p$. Determine the gain correction for the analog Chebyshev filter so that its passband response will vary between $1 + \delta_p$ and $1 - \delta_p$.

(c) Using your formula and expression from (a) and (b) above, determine the minimum value of N needed for the design convention in (b) assuming that $\delta_p = 0.02$. □

EXERCISE 3.1.14. **Chebyshev II Filters**

(a) Use the functions **f acheby2** and **f atransform** to design a fourth-order analog Chebyshev II lowpass filter with stopband cutoff frequency $\Omega_s = 5\pi = 15.708$ and $\epsilon = 0.05$. First design the fourth-order Chebyshev II filter with $\Omega_s = 1$, as occurs automatically using the function **f acheby2**. Display the pole/zero plot using **g apolezero**. Adjust the s-plane axis (downward) using *option 2* in the main menu. Sketch the pole/zero plot. The function **f atransform** can be used to perform a frequency scaling if its six parameters are specified as $c_0 = 0$, $c_1 = 1$, $c_2 = 0$, $d_0 = 5\pi$ (or 15.708), $d_1 = d_2 = 0$. Do this next.

(b) Display and sketch the magnitude, log magnitude, and phase responses using **g afreqres** with $\Omega_L = 0$ and $\Omega_H = 30$. If the passband is defined as those frequencies for which the passband error is less than 0.05, determine the passband cutoff frequency, Ω_p, for the filter. Use **g afilspec** for a more precise estimate. □

EXERCISE 3.1.15. **Determining the Design Parameters**

As we saw for Chebyshev I filters, the design parameters for an analog Chebyshev II filter are ϵ, N, and Ω_s. These are different from those that we have become used to. The purpose of this problem is to relate these parameters to the more familiar specifications.

(a) If the desired stopband ripple is Δ_s, determine the proper value of ϵ.

(b) Because the passband behavior of the filter is monotonic, the values of Δ_p and Ω_p are tightly coupled and both depend upon N. Using the properties of the Chebyshev polynomials that we have discussed, derive an approximate formula that relates Δ_p to Ω_p for fixed values of the other parameters. Test your formula by considering the design of a lowpass Chebyshev II filter with $N = 8$, $\Delta_s = 0.08$, $\Omega_p = 0.5$, and $\Omega_s = 1.0$. Determine Δ_p analytically and show your work. Verify your answer by using **f ocheby2**.

(c) Test your formula in part (b) by considering the design of a lowpass Chebyshev II filter with $\Delta_p = 0.01$, $\Delta_s = 0.08$, $\Omega_p = 0.5$, and $\Omega_s = 1.0$. Determine the order analytically and show your work. Verify your answer by using **f ocheby2**. □

EXERCISE 3.1.16. **Comparing Chebyshev Filters**

Design both a Chebyshev I and a Chebyshev II filter to satisfy the following specifications: $\Delta_p = 0.05$, $\Delta_s = 0.05$, $\Omega_p = 1$, $\Omega_s = 2$. You will need to make use of the **f ocheby1**, **f ocheby2**, **f acheby1**, **f acheby2**, and **f atransform** functions. Display and sketch the magnitude response and the group delay of each filter using **g afreqres** with $\Omega_L = 0$ and $\Omega_H = 3$. Which filter has the better group delay, i.e., the flatter group delay in the passband? Remember that the group delay in the stopband region is not important since the magnitude of the function in this region is approximately zero. □

EXERCISE 3.1.17. **Analog Elliptic Filters**

Elliptic filters are generally designed by computer programs that compute the coefficients. The function **f aelliptic** allows you to design analog elliptic filters. Use this function to design an analog filter based on the following specifications: $\Omega_s = 1.3$, $\Delta_p = 0.08$, $\Delta_s = 0.02$, and $N = 5$.

(a) Display the magnitude and log magnitude responses for frequencies between 0 and 2 rad/s using **g afilspec** and sketch the plots.

(b) Use **g afilspec** to determine the passband and stopband ripples, and passband and stopband cutoff frequencies (in rad/s) for the filter from your plots.

(c) Use **f obutter** and **f ocheby1** to determine the minimum orders required for a Butterworth and a Chebyshev I filter that meet the same specifications on Δ_p and Δ_s for the same cutoff frequencies as this elliptic filter? Observe that the elliptic filter achieves the same magnitude response characteristics with significantly lower order than the other filters. □

3.2 ANALOG FREQUENCY BAND TRANSFORMATIONS

The analog design functions that we have used up to this point—**f abutter**, **f abessel**, **f acheby1**, **f acheby2**, and **f aelliptic**—design only lowpass filters with a cutoff frequency of 1 rad/s, which can be varied by frequency scaling (see Exercise 3.1.6). Frequency scaling is a special case of a more general class of frequency transformations. Using these more general transformations, we can design highpass, bandpass, and bandstop filters directly from a lowpass prototype filter. These transformations are simple substitutions for the variable s in the system function that change the frequency selective characteristics of the filter. Some of them are summarized in Table 3.1. The lowpass-to-lowpass and lowpass-to-highpass transformations involve the parameters Ω_o and Ω_d, where Ω_o is the ideal cutoff frequency of the lowpass prototype and Ω_d is the corresponding cutoff frequency of the desired filter. For frequency transformations it is convenient to work with ideal cutoffs. This typically refers to the center cutoff frequency (i.e., the midpoint frequency in the transition band), but actually any frequency can be mapped.

For the lowpass-to-bandpass and lowpass-to-bandstop transformations, additional parameters are necessary because these filters have two cutoff frequencies. We have called these Ω_ℓ for the lower band edge and Ω_u for the upper band edge. For the case of the

Table 3.1. Frequency transformations for analog filters.

Type	Transformation
lowpass-to-lowpass	$s \rightarrow \dfrac{\Omega_o}{\Omega_d} s$
lowpass-to-highpass	$s \rightarrow \dfrac{\Omega_o \Omega_d}{s}$
lowpass-to-bandpass	$s \rightarrow \dfrac{\Omega_o s^2 + \Omega_o \Omega_\ell \Omega_u}{(\Omega_u - \Omega_\ell) s}$
lowpass-to-bandstop	$s \rightarrow \dfrac{\Omega_o (\Omega_u - \Omega_\ell) s}{s^2 + \Omega_u \Omega_\ell}$

lowpass-to-bandpass transformation, Ω_ℓ and Ω_u correspond to the ideal lower and upper cutoff frequencies of the passband whereas for the lowpass-to-bandstop transformation they correspond to the edges of the stopband. As before, Ω_o denotes the ideal cutoff frequency of the original lowpass prototype filter. The transition widths of the resulting filters are determined by the transition width of the lowpass prototype and by the specific transformation.

Performing these transformations by hand is tedious when the filter is of high order. However, they may be done easily by computer. The function **f atransform** performs the general transformation

$$s \rightarrow \frac{c_0 + c_1 s + c_2 s^2}{d_0 + d_1 s + d_2 s^2}.$$

It will ask you to specify values for $c_0, c_1, c_2, d_0, d_1,$ and d_2. By selecting these coefficients properly, all of the analog frequency transformations can be performed without much effort.

As an example of how these transforms work, consider the replacement of s by $1/s$. This is a simple example of a lowpass-to-highpass transformation with $\Omega_0 = \Omega_d = 1$,

$$H_{HP}(s) = H_{LP}(\frac{1}{s}).$$

Assume that $H_{LP}(s)$ is the Chebyshev lowpass filter shown in Figure 3.10. The frequency responses of the two filters are related by

$$H_{HP}(j\Omega) = H_{LP}(\frac{1}{j\Omega}).$$

Notice that at every frequency the value of $H_{HP}(j\Omega)$ is equal to the value of $H_{LP}(j\Omega)$ at some other frequency. Since $H_{LP}(j\Omega)$ has a passband, where the magnitude is nominally unity, and a stopband, where the value is nominally zero, $H_{HP}(j\Omega)$ has a passband and a stopband as well; only the frequencies covered by the bands are different. At $\Omega = 0$ the value of $H_{HP}(j\Omega)$ is the same as the value of $H_{LP}(j\Omega)$ at infinite frequency (i.e., $\Omega = \infty$) and vice versa. The value of the two filters at $\Omega = 1$ is the same. The filter $H_{HP}(j\Omega)$ is therefore a highpass filter with a cutoff frequency of 1 rad/s as shown in Figure 3.10b.

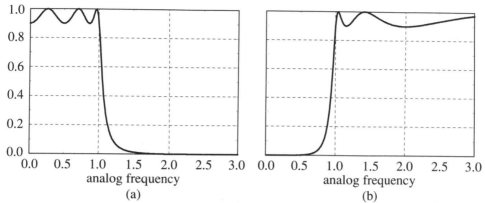

Figure 3.10. A lowpass-to-highpass transformation. (a) A Chebyshev lowpass filter. (b) The highpass filter that results by replacing s by $1/s$.

Analog frequency band transformations can be used in conjunction with the order estimation formulas. For example, a highpass filter that is derived from a frequency transformation of a lowpass prototype filter will have the same order. Since the highpass specifications can be converted to corresponding specifications for the lowpass prototype, the order estimation formulas discussed earlier can be used. Bandpass and bandstop filters are slightly more complicated. These may also be derived from lowpass prototypes, but the order of the bandpass or bandstop will be twice that of the prototype. Furthermore, bandpass filters have two transition bands instead of one. So which of the two transition band specifications should be substituted into the order formulas? One approach is to determine two lowpass prototypes and then design the one with the more restrictive specifications. Similarly, if we have a bandpass specification with different attenuations in the two stopbands, we must design for the tighter specification.

EXERCISE 3.2.1. Simple Application of the Transformations

The function **f atransform** can be used to perform any one of the transformations in Table 3.1. This exercise will use these transformations. Begin by designing a sixth-order Butterworth lowpass filter prototype, $H_a(s)$, using **f abutter**. This function will produce a lowpass filter with a cutoff frequency of 1. Display $H_a(s)$ using **g afreqres** with $\Omega_L = 0$ and $\Omega_H = 6$.

(a) Consider converting the prototype to a highpass filter with cutoff frequency equal to 2. This implies the transformation $s \rightarrow 2/s$. Use the function **f atransform** to realize this conversion. Note that by specifying $c_0 = 2$, $c_1 = 0$, $c_2 = 0$, $d_0 = 0$, $d_1 = 1$, and $d_2 = 0$, you will realize this transformation. Display the magnitude and log magnitude responses using **g afreqres** with $\Omega_L = 0$ and $\Omega_H = 6$ and sketch the plots.

(b) From part (a) the mechanics of performing analog frequency transformations should be clear. Following the same general procedure, convert the lowpass prototype to a bandpass filter with $\Omega_\ell = 2$ and $\Omega_u = 3$. Display the magnitude and log magnitude responses using **g afreqres** with $\Omega_L = 0$ and $\Omega_H = 6$ and sketch the plots.

(c) Now convert the prototype to a bandstop filter with $\Omega_\ell = 2$ and $\Omega_u = 3$. Display the magnitude and log magnitude responses using **g afreqres** $\Omega_L = 0$ and $\Omega_H = 6$ and sketch the plots. $\square$

EXERCISE 3.2.2. Designing a Butterworth Highpass Filter

The typical filter design problem is to design a particular frequency selective filter, such as a bandpass or highpass filter, that satisfies a given set of frequency domain specifications. Consider the design of the lowest-order Butterworth highpass filter, $H_a(s)$, that will meet the following specifications: $\Omega_s = 3$, $\Omega_p = 4$, $\Delta_s = 0.01$, $\Delta_p = 0.02$. To do this you should first determine the appropriate lowpass prototype specifications and then the appropriate highpass transformation parameter Ω_d. Record these specifications. Design the lowpass prototype and the highpass filter using **f obutter**, **f abutter**, and **f atransform**. Check your lowpass prototype characteristics with **g afilspec** after using **f obutter** and **f abutter** to verify that it satisfies the desired specifications. Since **f obutter** only provides an estimate of the order, it is possible that a lower order can also satisfy the specified constraints. Use **g afilspec** with $\Omega_L = 0$ and $\Omega_H = 6$ to display and sketch $|H(j\Omega)|$. $\square$

EXERCISE 3.2.3. A Chebyshev Bandpass Filter

For lowpass-to-lowpass and lowpass-to-highpass transformations, the orders of the resulting filters remain the same. However, this property is not true of the lowpass-to-bandpass and lowpass-to-bandstop transformations since these transformations involve substituting functions composed of second-order polynomials for s. In this exercise, you will face this issue as you attempt to design the eighth-order Chebyshev I bandpass filter, $H_a(s)$.

(a) Determine the order of the lowpass prototype needed to design an eighth-order bandpass filter.

(b) Use **f acheby1** with $\epsilon = 0.48$ to design the lowpass prototype. The bandpass filter, $H_a(s)$, is to have lower and upper passband cutoff frequencies of $\Omega_\ell = 1$ and $\Omega_u = 2$. Display the magnitude and log magnitude responses of your prototype filter using **g afreqres** with $\Omega_L = 0$ and $\Omega_H = 3$ and sketch these plots. Next design the bandpass filter using **f atransform**. Use **g apolezero** to display the pole/zero plot and sketch it. Rescale the s-plane axis so that all roots are within the field of view. $\square$

EXERCISE 3.2.4. Power Complementary Filters

The two filters with frequency responses $G_a(j\Omega)$ and $H_a(j\Omega)$ are said to be *power complementary* if

$$|G_a(j\Omega)|^2 + |H_a(j\Omega)|^2 = 1.$$

An ideal lowpass and highpass filter with the same cutoff frequency, for example, are power complementary. Power complementary filters are often used in the design of filter banks. This exercise looks at the analysis and design of power complementary filters.

(a) Show analytically that an analog Butterworth lowpass filter is the power complement of an analog Butterworth highpass filter if the cutoff frequencies are properly chosen.

How is the 3-dB cutoff frequency of the lowpass related to the 3-dB cutoff frequency of the highpass? How are the two filter orders related?

(b) Using **f abutter** and **f atransform**, design a pair of sixth-order power complementary Butterworth filters, $G_a(j\Omega)$ and $H_a(j\Omega)$. Display the magnitude response of $G_a(j\Omega)$ using **g afreqres**. Upon exiting the function, a file called _fr_ is created that contains the frequency response values that were plotted to the screen. Copy this to the file **temp1**. Now display the magnitude response of $H_a(j\Omega)$ using **g afreqres** and copy _fr_ into the file **temp2**. You may use the **g look** function to display the frequency responses in **temp1** and **temp2** to verify that they are correct.

(c) Next, compute the squared magnitude (or spectral power) $|G_a(j\Omega)|^2$ and $|H_a(j\Omega)|^2$ by typing

$$\textbf{f multiply} \quad \textbf{temp1} \quad \textbf{temp1} \quad \textbf{temp3}$$
$$\textbf{f mag} \quad \textbf{temp3} \quad \textbf{temp3}$$

and

$$\textbf{f multiply} \quad \textbf{temp2} \quad \textbf{temp2} \quad \textbf{temp4}$$
$$\textbf{f mag} \quad \textbf{temp4} \quad \textbf{temp4}$$

Using the **g slook2** function, display these squared magnitude responses superimposed on the same plot and sketch them. Now type

$$\textbf{f add} \quad \textbf{temp3} \quad \textbf{temp4} \quad \textbf{temp5}$$

This will add the squared responses together. Use the **g look** function to display **temp5**. If the power complementary filters were designed properly, the displayed result should have constant amplitude.

(d) Show that a Chebyshev Type I lowpass is the power complement of a Chebyshev Type II highpass. Relate the orders of the filters and their passband and stopband cutoff frequencies. How is ϵ of the lowpass related to ϵ of the highpass? Repeat the procedure outlined in part (c) for computing and displaying the squared magnitude responses. Sketch each plot as before.

(e) (optional) Repeat part (d) for a Chebyshev Type II lowpass and a Chebyshev Type I highpass. □

EXERCISE 3.2.5. Analog Highpass Design

Using **f ocheby2**, **f acheby2**, and **f atransform**, design an analog Chebyshev II highpass filter of minimum order to satisfy the following specifications: $\Omega_s = 4$, $\Omega_p = 6$, $\Delta_s = 0.01$, and $\Delta_p = 0.05$. Record the filter coefficients and provide a sketch of the filter's magnitude and log magnitude responses using **g afreqres** with $\Omega_L = 0$ and $\Omega_H = 12$. Carefully describe the steps taken to design the filter. □

EXERCISE 3.2.6.   **Analog Bandpass Design**

Using **f ocheby1**, **f acheby1**, and **f atransform**, design an analog Chebyshev I bandpass filter of minimum order to satisfy the following specifications: stopband from 0 to 3 rad/s with a maximum gain of 0.1; passband from 4 to 5 rad/s with a maximum error of 0.05; stopband from 6 to ∞ rad/s with a maximum gain of 0.1. Record the filter coefficients and provide a sketch of the filter's magnitude and log-magnitude responses using **g afreqres** with $\Omega_L = 0$ and $\Omega_H = 12$. Carefully describe the steps taken to design the filter. ☐

EXERCISE 3.2.7.   **Analog Bandstop Design**

Use **f aelliptic** and **f atransform** to design an analog elliptic bandstop filter of minimum order to satisfy the following specifications: passband from 0 to 2 rad/s with a maximum error of 0.01; stopband from 3 to 6 rad/s with a maximum gain of 0.01; passband above 7 rad/s with a maximum error of 0.01. Record the filter coefficients and provide a sketch of the filter's magnitude and log-magnitude responses using **g afreqres** $\Omega_L = 0$ and $\Omega_H = 12$. Carefully describe the steps taken to design the filter. ☐

3.3 ANALOG-TO-DIGITAL TRANSFORMATIONS

There are many techniques for converting analog filters (which we will refer to as *analog prototypes*) into digital filters. While many of these appear at first glance to be reasonable, some have proven to work well and some have not.

The motivation for designing digital filters by transforming analog prototypes is that the theory of analog filter design is well understood, and many convenient design formulas exist. For the approach to be useful, however, the transformation must exhibit several properties. First, if the analog filter has a rational system function in s, the digital filter must have a rational system function in z, preferably of the same order. The reason for this requirement is that all of our analog design techniques produce rational system functions and that digital filters with rational system functions are the only ones that we can implement using difference equations. Second, the frequency response of the digital filter must resemble the frequency response of the analog filter in some way in order to be useful. We neither require nor expect that the frequency responses will be identical; after all, the frequency response of a digital filter is periodic in frequency and the frequency response of an analog filter is not. Nonetheless, we would hope, for example, that a Butterworth analog lowpass filter would map to some type of monotonic lowpass digital filter. Otherwise, we would not be able to design digital filters to meet prescribed specifications. Finally, if the transformation is applied to a causal stable analog filter, the digital filter that results should also be causal and stable.

This section will look at several approaches for converting analog filters into digital ones. In each case, a sampling period parameter T appears in the transformation equation. It can be used to provide control over the cutoff frequencies or it can be set to one, in which case the analog frequency transformations in Table 3.1 can be used to provide this control.

3.3.1   Impulse Invariance

Impulse invariance is based on sampling. The impulse response of the digital filter, $h[n]$, is found by sampling the impulse response of the analog filter $h_a(t)$ and scaling the samples. If, for example, $h_a(t)$ is the impulse response of an analog filter, then

$$h[n] = Th_a(nT) \qquad (3.34)$$

is the impulse response of the digital filter. The sampling period, T, is a free parameter with this technique. The motivation behind the approach should be quite obvious. The sampling theorem says that a bandlimited signal, in this case a filter impulse response, can be sampled while preserving the shape of its Fourier transform, i.e., the filter's frequency response. The precise relationship between the frequency responses of the analog prototype filter, $H_a(j\Omega)$, and the digital filter, $H(e^{j\omega})$, is

$$H(e^{j\omega}) = \sum_{k=-\infty}^{\infty} H_a(j\frac{1}{T}(\omega + 2\pi k)). \qquad (3.35)$$

Notice that there is no amplitude scaling factor of $1/T$ in (3.35) as is usually the case for sampling. This is because $h_a(t)$ is normalized by the scaling factor T in (3.34) so that the peak magnitude of the digital filter's frequency response will be approximately equal to that of the analog prototype filter. Also notice that if $h_a(t)$ decays to zero with increasing t, as the impulse responses of stable filters do, then $h[n]$ will decay to zero also. It is, therefore, reasonable to hope that the digital filter will be stable. One of the major disadvantages of impulse invariance is that it can only be applied to analog filters that are bandlimited. Thus, it can be used to design lowpass and bandpass filters, but not highpass and bandstop filters. This is less of a disadvantage than it might appear, however, because we can always design a digital highpass filter by first designing a digital lowpass followed by a *digital* frequency transformation. These will be discussed shortly.

Impulse invariance is not a one-to-one mapping between the s-plane and the z-plane. Many points in the s-plane are mapped to the same point in the z-plane as illustrated in Figure 3.11. The precise relationship between $H_a(s)$ and $H(z)$ is

$$H(z) = \sum_{k=-\infty}^{\infty} H_a(\frac{1}{T}(\log|z| + j(\angle z + 2\pi k))), \qquad (3.36)$$

which follows from the sampling theorem. Each of the horizontal strips shown in Figure 3.11 is mapped to the interior of the z-plane unit circle. Notice that this means that all of the poles of the causal, stable analog filter will map to the interior of the unit circle, producing a causal, stable digital filter. The width of the strips is $2\pi/T$; T should be chosen to align the analog frequency axis with the discrete frequency axis. Ignoring the aliasing implied by (3.35), the analog and discrete frequencies are related by

$$\omega = \Omega T. \qquad (3.37)$$

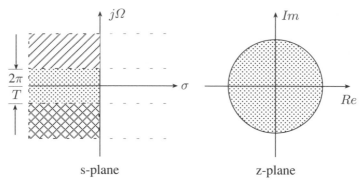

Figure 3.11.   Impulse invariance in the transform plane.

When $H_a(s)$ is rational, impulse invariance can be implemented very easily. To see this, consider the simple first-order system

$$H_a(s) = \frac{1}{s+a}.$$

Taking the inverse transform produces the result

$$h_a(t) = e^{-at}u(t).$$

Sampling the analog impulse response yields the digital one. Thus,

$$h[n] = Th_a(nT) = Te^{-anT}u[n].$$

Finally, the system function of the digital filter is

$$H(z) = \frac{T}{1 - e^{-aT}z^{-1}}.$$

The analog filter with a single pole at $-a$ maps to a digital filter with a single pole at e^{-aT}.

The general case is straightforward because sampling is a linear operation. Classical analog filters have distinct poles and thus may be expressed in a partial fractions expansion

$$H_a(s) = A_0 + \sum_{k=1}^{N} \frac{A_k}{s - s_k},$$

where the A_ks are constant coefficients. Applying the result from the one-pole example to the Nth-order analog prototype, we obtain the impulse invariant digital filter

$$H(z) = A_0 + \sum_{k=1}^{N} \frac{TA_k}{1 - e^{s_k T}z^{-1}}. \tag{3.38}$$

A pole of the analog filter at $s = s_k$ is mapped to a pole in the digital filter at $z = e^{s_k T}$, and the residues A_k are scaled by T. We see that zeros in the digital filter result from an interaction of the terms in the sum in (3.38) and do not result from mapping the analog zeros. Indeed the number of zeros for the two filters may not even be the same.

Equation (3.38) can be applied to all the analog prototypes that we have discussed in this chapter, because all of them have distinct poles. When the prototype analog filter has multiple poles at the same location, the procedure needs to be modified slightly. For this case, consider the analog filter with a double pole at $s = -a$:

$$H_a(s) = \frac{1}{(s+a)^2}.$$

Following the procedure above for the single pole case, the digital filter that results by impulse invariance is easily seen to be

$$H(z) = \frac{T^2 e^{-aT} z^{-1}}{(1 - e^{-aT} z^{-1})^2}.$$

Similar expressions can be derived for higher-order poles.

The residues and the poles in the partial fractions expansion above are usually complex numbers. We can recast the procedure so that only real valued residues are involved by using second-order sections instead of first-order sections. The impulse invariance transformation for the two types of proper second-order sections are

$$\frac{a}{s^2 + 2bs + a^2 + b^2} \quad \rightarrow \quad \frac{T z^{-1} e^{-bT} \sin aT}{1 - 2 z^{-1} e^{-bT} \cos aT + z^{-2} e^{-2bT}} \tag{3.39}$$

$$\frac{s + b}{s^2 + 2bs + a^2 + b^2} \quad \rightarrow \quad \frac{T(1 - z^{-1} e^{-bT} \cos aT)}{1 - 2 z^{-1} e^{-bT} \cos aT + z^{-2} e^{-2bT}}. \tag{3.40}$$

Any second-order filter with complex poles can be transformed using one or the other of these formulas or by a linear combination of the two.

In summary, the impulse invariance technique for the design of digital filters can be accomplished through the application of the following steps:

1. Design an appropriate bandlimited analog filter.
2. Split it into the sum of first- and second-order analog sections and transform each section into a first- or second-order digital section using (3.38), (3.39), and (3.40).
3. Manipulate the digital system function into the desired implementation form (e.g. direct form, cascade of sections, etc.). This latter topic is discussed more fully in Chapter 5.

Performing these impulse invariance manipulations manually can be very tedious; however, by using a computer, the operation can be done easily. The function **f impinv** performs this operation and is available for your use in the following exercises. The function **f impinv** is restricted to analog filters with distinct roots, but this restriction should not pose a problem since all of the classical analog filters have distinct roots.

EXERCISE 3.3.1. **Applying the Impulse Invariance Method**

The impulse invariance method transforms an analog filter into a digital one as discussed in the introduction. A convenient way to perform this transformation is to first convert the analog filter into the sum of first-order sections and then perform the transformation using (3.38).

To illustrate this process, consider the analog filter

$$H_a(s) = \frac{1}{s+1+j} + \frac{1}{s+1-j} - j\frac{1}{s+2+2j} + j\frac{1}{s+2-2j}.$$

(a) Without the aid of the computer, apply (3.38) with $T = 1$ and convert $H_a(s)$ into a digital filter, $H(z)$. Note that $H(z)$ has complex coefficients. Using the *create file* option in **f siggen** create files for each of the four first-order terms in $H(z)$ with a zero starting point. Next use the function **f par** to convert $H(z)$ to direct form. Using **g polezero**, display and sketch the pole/zero plot of its output file.

(b) In part (a), you observed that the result of the transformation based on first-order sections led to complex filter coefficients. Filters with complex coefficients require more multiplication and addition operations than filters with real coefficients. If the first-order sections each have their corresponding complex conjugate section present, as is usually the case (and which is the case in our example), the analog filter can be expressed as a sum of second-order sections, each with real coefficients. Without the aid of a computer, convert $H_a(s)$ into second-order sections with real coefficients and apply (3.38) (with $T = 1$) to obtain $H(z)$. Use **f siggen** to construct signal files for the two second-order sections, the **f par** function to convert the second-order sections to direct form, and **g polezero** to display the pole/zero plot. Sketch this plot in your write-up.

(c) The function **f impinv** will take an analog filter in direct form and convert it to a direct-form digital filter using the impulse invariance method. It does this following the same procedure discussed in part (a). Design a sixth-order elliptic filter using **f elliptic** with $\Delta_p = 0.01$, $\Delta_s = 0.001$, and $\Omega_p = 2$. Display the analog pole/zero plot and magnitude response, using **g apolezero**. Next use the function **f impinv** with $T = 1$ to convert $H_a(s)$ to digital form directly. Display and sketch the pole/zero plot and magnitude response using **g polezero**. Use the *list pole/zero* submenu option to list the zeros and record these values. Notice that some of them are very far away from the unit circle and thus have little effect on the quality of the filter.   □

EXERCISE 3.3.2. **Choosing the Analog Design Parameters**

The procedure as presented for designing a digital IIR filter begins by designing an analog filter with the proper specifications and then using impulse invariance to obtain $H(z)$. Consider the design of a digital elliptic filter with passband cutoff $\omega_p = \pi/4$, stopband cutoff $\omega_s = \pi/3$, passband ripple $\Delta_p = 0.01$, and stopband ripple $\Delta_s = 0.01$. Assume that you have available an analog filter design program that allows you to specify the cutoff frequencies Ω_p and Ω_s. This exercise does not require the use of the computer.

(a) What analog cutoff frequencies (Ω_p, Ω_s) should be selected for the analog prototype if $T = 34524$ in the impulse invariance method.

(b) What value of T should be chosen if the analog prototype has cutoff frequencies $\Omega_p = 6\pi$ and $\Omega_s = 8\pi$?

(c) The function **f aelliptic** provided in the software constrains the passband cutoff frequency Ω_p to be unity. The stopband cutoff frequency Ω_s is a free parameter that is greater than one. Determine Ω_s and T so that a digital elliptic filter can be designed to meet the digital passband and stopband cutoff constraints given in the introduction to the problem. Verify your answer by designing the digital elliptic filter using **f aelliptic** and **f impinv** and examining its magnitude response characteristics using **g dfilspec**.   □

EXERCISE 3.3.3.   Application of Impulse Invariance to LPFs and HPFs

Using **f abutter**, design a fifth-order Butterworth filter, $H_a(s)$.

(a) Use the function **f impinv** to determine a digital Butterworth filter using impulse invariance with $T = 0.5$. Display and sketch the magnitude and log magnitude responses of the digital filter using **g dtft**.

(b) Now use **f atransform** to convert $H_a(s)$ into a highpass filter with cutoff frequency $\Omega_d = 1$. Note that this can be accomplished by specifying $c_0 = 1$, $c_1 = 0$, $c_2 = 0$, $d_0 = 0$, $d_1 = 1$, and $d_2 = 0$. Display the frequency response of the analog highpass filter using **g afreqres** with $\Omega_L = 0$ and $\Omega_H = 3$ and verify that it works properly. Apply the impulse invariance method as in part (a) to convert this highpass filter to a digital one. Display the magnitude response using **g dtft** and sketch the plots. Explain why the digital filter has poor spectral characteristics.   □

EXERCISE 3.3.4.   Impulse Invariance of LPFs

Using **f ocheby1** to estimate the design parameters and **f acheby1**, design a sixth-order Chebyshev lowpass filter $H_a(s)$ with $\Omega_p = 1$, $\Omega_s = 2$, and $\Delta_p = \Delta_s = 0.01$.

(a) Use **f impinv** to determine the corresponding digital filter resulting from the impulse invariance transformation with $T = 0.5$. Display and sketch the magnitude and log magnitude responses of this filter using **g dtft**.

(b) Now repeat part (a) with $T = 2$ and again with $T = 3$. Display the magnitude responses for both cases using **g dtft** and sketch the plots. Explain why the digital filter is not a good lowpass filter in the case where $T = 3$.   □

EXERCISE 3.3.5.   Designing Filters Using Impulse Invariance

In the previous exercises, you started with design specifications for the analog filter and converted the filter to digital form using impulse invariance. Now consider starting with a set of digital lowpass filter specifications and designing the digital filters.

Assume we wish to design digital Butterworth, Chebyshev, and elliptic filters to meet the following digital specifications: $0.9 \leq |H(e^{j\omega})| \leq 1$ in the passband, $|H(e^{j\omega})| \leq 0.01$ in the stopband, $\omega_p = \pi/2 = 1.5708$, and $\omega_s = 3\pi/4 = 2.356$. For each filter, record the order and provide a sketch of its magnitude response. The function **g dfilspec** can be used to display the magnitude responses.

(a) If the digital cutoff frequencies are ω_p and ω_s, what are the corresponding analog cutoff frequencies Ω_p and Ω_s, assuming that impulse invariance is used. Note that the expressions for Ω_p and Ω_s are functions of T. What value of T should be used if Ω_p is constrained to equal 1.

(b) Design a digital elliptic lowpass filter, $H_e(z)$, of lowest order that satisfies the specifications above using **f aelliptic** and **f impinv**. Check the magnitude response of $H_e(z)$ using **g dfilspec** and verify that it meets the above-stated criteria. If not, redesign it so that the specifications are met.

(c) Design a digital Chebyshev I lowpass filter, $H_c(z)$, of lowest order that satisfies the specifications above using **f ocheby1**, **f acheby1**, and **f impinv**. Verify that $H_c(z)$ satisfies the design criteria using **g dfilspec**.

(d) Design a digital Butterworth lowpass filter, $H_B(z)$, of lowest order that satisfies the specifications above. Verify that $H_B(z)$ satisfies the design criteria using **g dfilspec**. Pay attention to the fact that Ω_p is less than 1 in the Butterworth design function because the 3-dB cutoff frequency is unity. □

EXERCISE 3.3.6. **Step Invariance**

Step invariance is a different method for mapping an analog prototype into a digital filter that is closely related to the impulse invariance approach. With impulse invariance the impulse responses of the analog and digital filters are related by sampling. With a step invariance design, the step responses of the two filters are related in this fashion. The step response of an analog filter is $s_a(t)$, where

$$s_a(t) = \int_{-\infty}^{t} h_a(\tau)d\tau.$$

This leads to

$$S_a(j\Omega) = \frac{1}{j\Omega}H_a(j\Omega)$$

and

$$S_a(s) = \frac{1}{s}H_a(s)$$

in the Fourier and Laplace domains, respectively. In the z-domain, the z-transform of the step response is related to the system function by

$$S(z) = \frac{1}{1 - z^{-1}}H(z).$$

Because sampling is applied to the step response rather than to the impulse response, $S_a(s)$ and $S(z)$ are related by the same transformations that related $H_a(s)$ to $H(z)$ with the impulse invariance technique (3.38).

(a) Consider the design of a digital Butterworth filter using the step invariance method. Design a fourth-order Butterworth filter $H_a(s)$ using **f abutter**. Next compute the step response for the Butterworth filter $S_a(s)$. This can be done using **g apolezero** by adding a pole to $H_a(s)$ at $s = 0$. However, since a pole at zero in the s-plane corresponds to an unstable filter and consequently may cause overflow problems in the software, specify the pole to be at $s = -0.01$. Use the *write to output file* option to put the step response filter in a file. Use the function **f impinv** with $T = 0.5$ to convert $S_a(s)$ to the digital filter $S(z)$. The last step is to convert $S(z)$ to $H(z)$ via the equation

$$H(z) = (1 - z^{-1})S(z).$$

Ideally, this would involve introducing a zero at $z = +1$, which in turn would remove the pole at $z = 1$. Since our pole and, consequently, our corresponding zero as well, only have locations approximately equal to $z = 1$, it is more convenient to use the *change pole/zero* option in the function **g polezero** to remove this pole. Do this and write the new filter to a file using the *write to output file* menu option.

(b) Now apply impulse invariance to $H_a(s)$ using **f impinv** with $T = 0.5$. Display the magnitude response of this digital filter using **g dtft**. Ignoring any differences in the gain, how do the two filters compare? Is one clearly better than the other or are they about the same?

(c) Design a fourth-order elliptic filter with $\Delta_p = \Delta_s = 0.05$ and $\Omega_s = 1.5$. Repeat the step invariance design procedure outlined in the previous parts of the exercise and compare the results with that of impulse invariance. Use the **g polezero** function to display and sketch the pole/zero and magnitude response plots. Comment on any differences you observe with respect to the two filters. $\square$

3.3.2   The Bilinear Transformation

The bilinear transformation is an invertible nonlinear mapping between the s-plane and the z-plane defined by

$$s = m(z) = \beta \frac{1 - z^{-1}}{1 + z^{-1}}$$

or

$$z = m^{-1}(s) = \frac{1 + \dfrac{s}{\beta}}{1 - \dfrac{s}{\beta}},$$

where β is a free parameter that, historically, has been set equal to $2/T$, although T does not have a useful interpretation as a sampling interval. A digital filter designed by the bilinear transformation is related to its analog prototype by

$$H(z) = H_a(m(z)).$$

The transformation preserves both the order and the stability of the prototype. Points in the left-half of the s-plane are mapped to the interior of the z-plane unit circle, and points on the $j\Omega$-axis are mapped nonlinearly onto the unit circle itself. This latter mapping is given by

$$\Omega = \beta \tan \frac{\omega}{2} = \frac{2}{T} \tan \frac{\omega}{2} \qquad (3.41)$$

or equivalently

$$\omega = 2 \tan^{-1} \frac{\Omega}{\beta} = 2 \tan^{-1} \frac{\Omega T}{2}. \qquad (3.42)$$

The analog frequency origin is mapped to the digital origin, and analog frequencies at $\pm\infty$ are mapped to $z = -1$.

The bilinear transformation is applied in the same way as the analog frequency transformations, i.e., by substituting the transformation function into $H_a(s)$ for every occurrence of s. The function **f bilinear** is provided to allow you to perform the transformation easily. By varying the value of β, the ideal cutoff frequency of the filter may be changed. This degree of freedom can be used to design digital filters with arbitrary cutoff frequencies from analog prototypes with fixed cutoffs (such as those with $\Omega_p = 1$ as produced by many of our analog prototype routines).

The function **f bilinear** also allows for arbitrary custom substitutions of the form

$$s \rightarrow \frac{c_0 + c_1 z^{-1} + c_2 z^{-2}}{d_0 + d_1 z^{-1} + d_2 z^{-2}}$$

for analog-to-digital transformations and transforms of the form

$$z \rightarrow \frac{c_0 + c_1 s^1 + c_2 s^2}{d_0 + d_1 s^1 + d_2 s^2}$$

for digital-to-analog substitutions.

The bilinear transform is a popular technique for constant band filters, such as the ones that we have considered in this chapter. Digital elliptic filters obtained by mapping analog elliptic filters are optimal in the same sense that analog elliptic filters are optimal. The nonlinear frequency mapping, however, may be undesirable for filters where the frequency response is nonconstant or for filters where the group delay response is particularly important, such as Bessel filters.

EXERCISE 3.3.7. Digital Elliptic Filter

An analog elliptic filter is equiripple. Here we examine the ripples of digital elliptic filters designed by impulse invariance and the bilinear transformation.

Design a sixth-order analog elliptic filter, $H_a(s)$, with $\Omega_p = 1$, $\Omega_s = 1.5$, $\Delta_p = 0.001$, and $\Delta_s = 0.01$.

(a) Use **g afilspec** with $\Omega_L = 0$ and $\Omega_H = 3$ to measure the passband and stopband ripples of the filter. Sketch the magnitude response of the filter.

(b) Convert this prototype to a digital filter using impulse invariance with $T = 0.5$. Display and sketch the magnitude response of the digital filter using **g dfilspec** and measure the passband and stopband cutoff frequencies. Is the filter equiripple? How many ripples are present in the stopband?

(c) Repeat (b) using **f bilinear**, which implements the bilinear transform, with $\beta = 2$. Compare the number of ripples and ripple amplitudes to those of $H_a(s)$.

What do you conclude about the preservation of the equiripple nature of the analog elliptic filter after undergoing impulse invariance and bilinear transformation?   □

EXERCISE 3.3.8.   Digital Bessel Filters

The purpose of this exercise is to compare digital Bessel filters designed with the impulse invariant transformation with ones designed with the bilinear transformation.

(a) Design a sixth-order analog Bessel filter using **f abessel**. Display and sketch its magnitude and group delay responses using **g afreqres** with $\Omega_L = 0$ and $\Omega_H = 20$.

(b) Convert the analog filter in part (a) to a digital filter using **f impinv** with $T = 1/6$. Display and sketch the magnitude and group delay of the digital filter using **g dtft**.

(c) Now convert the analog filter in part (a) to a digital filter using **f bilinear** with $\beta = 4.0$. Display and sketch the magnitude and group delay of the resulting filter as you did in part (b).

(d) Of the two digital filters that you designed, which has the flatter group delay in the passband region $-\pi/4 \leq \omega \leq \pi/4$?   □

EXERCISE 3.3.9.   The Effect of β in the Bilinear Transformation

The bilinear transformation has a free parameter β, which, by convention, is often set to 1 or 2. The purpose of this exercise is to determine the effect of varying β on the design process.

Design a sixth-order analog Chebyshev I lowpass filter, $H_a(s)$, with $\Omega_s = 2$ and $\epsilon = 0.3$. Use the function **f bilinear** with $\beta = 2$ to convert the analog filter into a digital one. Display its magnitude response using **g dtft**.

(a) Design three digital filters using **f bilinear** applied to the analog prototype $H_a(s)$: the first with $\beta = 2$, the second with $\beta = 5$, and the third with $\beta = 10$. Display the magnitude responses using **g dtft** and sketch the plots. What effect does β have on the digital filter?

(b) Examine the formula for the bilinear transform and the formula for analog lowpass-to-lowpass conversion in Table 3.1. Show how β can be used to control the cutoff frequency in the bilinear transformation.

If analog filter design programs are available that allow for specification of arbitrary cutoff frequencies, the conventional bilinear transformation with $\beta = 1$ is convenient to use. If the analog design programs are restricted to have a unity cutoff frequency, it is often convenient to use the β parameter in the bilinear transformation as a control parameter for the cutoff frequency.   □

EXERCISE 3.3.10.   Filter Cutoff Frequencies for Analog Butterworth Filters

Normally, the frequency cutoff specifications for a digital lowpass filter are given by ω_p and ω_s. But for filters with maximally flat magnitude responses it is a common practice to specify the 3-dB cutoff frequency. Consider the analog 3-dB and digital 3-dB cutoff frequencies when the bilinear transform is used. The mappings between the two frequency variables are given in (3.42) and (3.41).

(a) If $\beta = 1$, determine the 3-dB cutoff frequency, Ω_c, for the analog Butterworth lowpass filter needed to design a digital Butterworth filter with 3-dB cutoff frequency $\omega_c = \pi/6$ using the bilinear transformation.

(b) Design an eighth-order analog prototype lowpass Butterworth filter using **f abutter** and **f atransform** that will have the cutoff frequency that you determined in (a). Determine the digital filter using **f bilinear**. Display the magnitude response of the digital filter using **g dfilspec** and sketch the plot.

(c) Performing an analog lowpass-to-lowpass translation can be avoided by choosing an appropriate value for β in the bilinear transform. Determine β and use it in the function **f bilinear** to design the digital filter directly from the unscaled Butterworth filter with $\Omega_c = 1$. Verify the correctness of your result by examining the magnitude response using **g dfilspec**. □

EXERCISE 3.3.11. Designing Filters to Meet Specifications

Because of the nonlinear frequency warping introduced by the bilinear transformation, the digital filter specifications must be prewarped to analog specifications before the analog filter is designed. In this problem, we want to design a digital Chebyshev I lowpass filter, $H(z)$, with $1 - \Delta_p \leq |H(\omega)| \leq 1$ in the passband, where $\Delta_p \leq 0.02$, $|H(\omega)| \leq \Delta_s$ in the stopband, and where $\Delta_s \leq 0.02$, $\omega_p = 0.2\pi = 0.6283$, and $\omega_s = 0.4\pi = 1.2566$.

(a) In order to design the analog prototype, $H_a(s)$, using the software, you must use **f acheby1**. As you recall, **f acheby1** is constrained to have $\Omega_p = 1$. Determine the value of β that should be used in **f bilinear** so that the transformed filter will meet the digital passband frequency specifications. Using this value of β, next determine the parameters of the analog prototype, $H_a(s)$, i.e., Ω_s, Δ_p, and Δ_s. Use **f ocheby1** to determine the order.

(b) Design the filter and then apply the bilinear transformation to $H_a(s)$ using **f bilinear**. Plot and sketch the magnitude and log magnitude responses of the digital filter using **g dfilspec**. Use the right and left arrow keys to verify that the filter meets the digital specifications.

(c) Using the procedure outlined above as a model, design a digital Butterworth and a digital elliptic lowpass filter to meet the following specifications: $\Delta_p = 0.02$, $\Delta_s = 0.001$, $\omega_p = 0.4\pi = 1.2566$, and $\omega_s = 0.7\pi = 2.199$. Record the orders for your filters. Display the log magnitude responses for these filters using **g dtft** and sketch the plots. □

EXERCISE 3.3.12. Optimality of Digital Chebyshev I Filters

An analog Chebyshev I filter is an optimal all-pole filter. Consider the design of an analog third-order lowpass Chebyshev I filter, $H_a(s)$, with stopband cutoff frequency $\Omega_s = 2.5$, a maximum passband ripple of 0.05, and a maximum stopband deviation of 0.1. Use **f ocheby1** and **f acheby1** to design $H_a(s)$. Next use **f bilinear** with $\beta = 2$ to convert $H_a(s)$ into the digital filter $H(z)$. Display the pole/zero plot using **g polezero** and sketch the pole/zero plot and magnitude response. Use the *list pole/zero* option to list the poles and zeros of the filter explicitly. Record the values of the zeros. Explain why the all-pole analog filter $H_a(s)$ does not become an all-pole digital filter. Why are the zeros of $H(z)$ located where they are?   □

EXERCISE 3.3.13. Other *s*-Plane to *z*-Plane Transformations

The bilinear transform is the most popular method for conversion of analog prototypes into digital filters. It has been examined extensively in many different disciplines and from many different viewpoints. One such viewpoint arises from relating the z-transform to the Laplace transform variable

$$z = e^{sT},$$

which can be expressed equivalently as

$$s = \frac{1}{T} \ln z.$$

The logarithm of z has several well-known power series expansions, three of which are:

$$\ln z = 2 \frac{z-1}{z+1} \left[1 + \frac{1}{3} \left(\frac{z-1}{z+1} \right)^2 + \frac{1}{5} \left(\frac{z-1}{z+1} \right)^4 + \frac{1}{7} \left(\frac{z-1}{z+1} \right)^6 \cdots \right], \quad (3.43)$$

$$\ln z = (z-1) \left[1 + \frac{1}{2}(1-z) + \frac{1}{3}(1-z)^2 + \frac{1}{4}(1-z)^3 + \ldots \right], \quad (3.44)$$

$$\ln z = 2 \frac{z-1}{z} \left[1 + \frac{1}{2} \left(\frac{z-1}{z} \right) + \frac{1}{3} \left(\frac{z-1}{z} \right)^2 + \frac{1}{4} \left(\frac{z-1}{z} \right)^3 \cdots \right]. \quad (3.45)$$

If we truncate the series expansion in (3.43) to one term, we obtain $2(z-1)/(z+1)$, which we recognize as the bilinear transformation with $\beta = 2$. In this exercise, we consider other transformations based on truncating the power series expressions in (3.44) and (3.45). First design a fifth-order analog elliptic lowpass prototype filter, $H_a(s)$, with $\Omega_s = 1.5$, $\Delta_p = 0.02$, and $\Delta_s = 0.01$ using **f aelliptic**. Now consider the following *s*-to-*z* transforms:

1. $s = z - 1 = \dfrac{1 - z^{-1}}{z^{-1}}$,

2. $s = -(1/2)z^2 + 2z - 3/2 = \dfrac{-0.5 + 2z^{-1} - 1.5z^{-2}}{z^{-2}}$,

3. $s = (z - 1)/z = 1 - z^{-1}$,

which were derived from the first term of (3.44), the first two terms of (3.44), and the first term of (3.45), respectively. Using these transforms, the analog prototype, $H_a(s)$, and the function **f bilinear**, design three digital filters $H_1(z)$, $H_2(z)$, and $H_3(z)$ by applying the transformations given above. For each display and sketch the pole/zero and magnitude response plots using **g polezero**. Do the same for the filter designed using the bilinear transform with $\beta = 2$. How do these filters compare to each other in terms of their magnitude response characteristics?

The bilinear transform preserves stability. That is, a stable, causal analog prototype will always be transformed into a stable, causal digital filter. Does this property hold for the transforms introduced above?   □

3.3.3 The Matched z-Transform

The bilinear transformation and impulse invariance are the most commonly used transformations for mapping analog prototypes to digital filters, but other mappings exist. The matched z-transform is one of these. If the analog system function is written in factored form as

$$H_a(s) = \frac{\displaystyle\prod_{k=1}^{M}(s - z_k)}{\displaystyle\prod_{k=1}^{N}(s - p_k)},$$

the matched z-transform is defined by the mappings

$$s - z_k \implies 1 - e^{z_k T} z^{-1}$$
$$s - p_k \implies 1 - e^{p_k T} z^{-1},$$

where T is the sampling period. Equivalently, the numerator and denominator factors of $H_a(s)$ can be expressed in terms of second-order factors, on each of which a mapping can be performed using the rule

$$(s + a)^2 + b^2 \implies 1 - 2e^{-aT} \cos bT z^{-1} + e^{-2aT} z^{-2}.$$

Notice that there are similarities between the matched z-transform and impulse invariance. The same mapping is used to map the poles, and there is a dependence on a sampling period. As might be expected, the aliasing issue that was discussed in the context of impulse invariance is also present here. Thus, in order to map an analog filter into the digital domain while preserving its spectral magnitude characteristics, severe aliasing should be avoided. The zeros of many analog filters are located on the $j\Omega$-axis. These can be used to determine a sampling period T that will result in negligible aliasing. As a rule of thumb, examine the highest frequency zero of the analog filter on the $j\Omega$-axis and treat it as a Nyquist frequency (or highest frequency component) in the filter. This can then be used to determine the sampling period T. For example, if the zero of $H_a(s)$ located on the $j\Omega$-axis farthest from the origin is 2.5 rad/s, then from sampling theory

$$\Omega_s = 2\pi f_s = \frac{2\pi}{T} > 2(2.5),$$

where Ω_s is the sampling frequency in rad/s and f_s is the sampling frequency in Hertz. Since we must sample at a rate above twice the Nyquist frequency, T should be less than $\pi/2.5$. (For filters that have no zeros, see Exercise 3.3.15.) A digital frequency transformation (see the next section) can then be used to adjust the cutoff frequency of the filter.

EXERCISE 3.3.14.   The Effect of Aliasing in the Matched z-Transform

Use **f aelliptic** to design a sixth-order analog elliptic filter, $H_a(s)$ with $\Delta_p = 0.01$, $\Delta_s = 0.01$, and $\Omega_s = 1.5$. Examine the pole/zero plot of $H_a(s)$ using **g apolezero** and sketch the pole/zero plot. Also examine the zeros using the *list pole/zero* option and record the value of the highest frequency zero on the $j\Omega$-axis. Based on your findings, what is the minimum sampling period T that should be used to avoid significant aliasing. Next use **f matchedz** to design the following four filters:

(a) $G_1(z)$ with $T = 0.3$,

(b) $G_2(z)$ with $T = 0.5$,

(c) $G_3(z)$ with $T = 0.8$, and

(d) $G_4(z)$ with $T = 1.0$.

Using **g polezero**, display and sketch the pole/zero plots for each of the previous digital filters. Use **g dtft** to display the magnitude and log magnitude responses for each digital filter.[5] Which values of T worked well and which did not?   □

EXERCISE 3.3.15.   Matched z-Transform with Allpole Filters

The rule of thumb suggested in the section introduction applied to filters with zeros. Consider now the application of the matched z-transform to all-pole filters.

Design an eighth-order Chebyshev I filter, $H_a(s)$, with $\epsilon = 0.2$ and an eighth-order Butterworth filter $G_a(s)$, using **f acheby1** and **f abutter**, respectively.

(a) Using **g apolezero**, display and sketch the pole/zero plots for $H_a(s)$ and $G_a(s)$. Use the *list pole/zero* option to carefully label your plot.

(b) Choose three values of T in the range 0.5 to 3 for use in the matched z-transform. Apply **f matchedz** with these values of T to $H_a(s)$ and $G_a(s)$. Using **g polezero**, display and sketch the resulting pole/zero and magnitude response plots for each digital filter.

(c) Explain why the magnitude response characteristics are so degraded for all-pole filters.   □

EXERCISE 3.3.16.   Matched z-Transform or Bilinear Transformation?

The bilinear transformation is almost always preferred over the matched z-transform; in fact, the matched z-transform is not used very often. Here we will see the general superior performance of the bilinear transform over the matched z-transform.

(a) Design a fifth-order analog elliptic filter with $\Omega_s = 1.5$ and $\Delta_p = \Delta_s = 0.01$, using **f aelliptic**. Convert the analog filter to a digital filter $H_1(z)$ using **f bilinear** with $\beta = 2$. Next, construct a second digital filter $H_2(z)$ using **f matchedz** with $T = 0.9$. Display and sketch the magnitude and log magnitude response plots for both filters. Use **g**

---

[5]For filters with many roots closely clustered together, the magnitude response subroutine in **g** polezero (which was written for fast execution) may produce erroneous magnitude response plots. If such behavior is suspected, the magnitude response should be computed and displayed using **g** dtft.

dfilspec to determine the cutoff frequencies and stopband and passband deviations precisely. Which digital filter has better overall characteristics in terms of a narrow transition width and small stopband and passband ripple?

(b) Design a sixth-order analog Bessel filter using **f abessel**. Display and sketch its magnitude and group delay response using **g afreqres** with $\Omega_L = 0$ and $\Omega_H = 9.0$.

Now convert this filter into two digital filters, as described in part (a) via the bilinear transformation and the matched z-transform. Plot and sketch the magnitude response and group delay of the digital filters using **g dtft**.

You will observe that the magnitude response for the matched z-transformed filter is very distorted. One way to improve these characteristics is to add zeros in the z-plane at $z = -1$. Use the *change pole/zero* option in **g polezero** to add zeros at $z = -1$ and sketch the new magnitude responses. Comment on the effectiveness of this method for improving the magnitude response characteristics of the filter.

How do the group delay characteristics of this filter compare to those of the filter designed by the bilinear transform? How do they compare to those of the analog Bessel filter?   □

3.4   DIGITAL FREQUENCY BAND TRANSFORMATIONS

Digital lowpass filters can be converted into other frequency selective filters such as highpass, bandpass, and bandstop filters by employing frequency band transformations in the z-domain. These transformations work much like the analog transformations that were discussed earlier.

We transform a lowpass filter with an ideal cutoff frequency of θ_c into another lowpass with its cutoff at ω_c using the substitution

$$z^{-1} \rightarrow \frac{z^{-1} - \alpha}{1 - \alpha z^{-1}},$$

where

$$\alpha = \frac{\sin\left(\frac{\theta_c - \omega_c}{2}\right)}{\sin\left(\frac{\theta_c + \omega_c}{2}\right)}.$$

It is often most useful to view the frequencies θ_c and ω_c as center cutoff or ideal cutoff frequencies. However, they can also assume the role of passband or stopband edge cutoff frequencies or 3-dB frequencies.

Similarly, a lowpass-to-highpass transformation can be performed using the substitution

$$z^{-1} \rightarrow -\frac{z^{-1} + \alpha}{1 + \alpha z^{-1}},$$

where

$$\alpha = -\frac{\cos\left(\dfrac{\theta_c + \omega_c}{2}\right)}{\cos\left(\dfrac{\theta_c - \omega_c}{2}\right)}.$$

A lowpass-to-bandpass transformation is effected with the transformation

$$z^{-1} \rightarrow -\frac{z^{-2} - \dfrac{2\alpha k}{k+1} z^{-1} + \dfrac{k-1}{k+1}}{\dfrac{k-1}{k+1} z^{-2} - \dfrac{2\alpha k}{k+1} z^{-1} + 1},$$

where

$$\alpha = \frac{\cos\left(\dfrac{\omega_{c2} + \omega_{c1}}{2}\right)}{\cos\left(\dfrac{\omega_{c2} - \omega_{c1}}{2}\right)}$$

$$k = \cot\left(\frac{\omega_{c2} - \omega_{c1}}{2}\right) \tan\left(\frac{\theta_c}{2}\right).$$

Finally, a lowpass-to-bandstop transformation can be accomplished using

$$z^{-1} \rightarrow \frac{z^{-2} - \dfrac{2\alpha}{1+k} z^{-1} + \dfrac{1-k}{1+k}}{\dfrac{1-k}{1+k} z^{-2} - \dfrac{2\alpha}{1+k} z^{-1} + 1},$$

where

$$\alpha = \frac{\cos\left(\dfrac{\omega_{c2} + \omega_{c1}}{2}\right)}{\cos\left(\dfrac{\omega_{c2} - \omega_{c1}}{2}\right)}$$

$$k = \tan\left(\frac{\omega_{c2} - \omega_{c1}}{2}\right) \tan\left(\frac{\theta_c}{2}\right).$$

For the bandpass and bandstop cases, ω_{c1} and ω_{c2} correspond to the lower and upper ideal cutoff frequencies for the bandpass passband and bandstop stopband, respectively. Note that as was the case for the analog frequency transformations, the width of the transition region cannot be controlled through these transformations, only the locations of cutoff frequencies. Transition widths must be addressed by properly designing the digital lowpass prototype.

EXERCISE 3.4.1. **Simple Application of the Digital Transformations**

The function **f dtransform** can be used to perform any one of the transformations given in this subsection. In this exercise, you will simply apply these transformations to a digital filter and observe how they work. Let $H(z)$ be the digital filter stored in the file **h2df**. Use **g dfilspec** to display the magnitude response and sketch the plot. Observe that the center cutoff frequency is 1.97 rad.

(a) Apply the digital transformations to convert $H(z)$ to a highpass filter with cutoff frequency $\pi/2 = 1.5708$. Use the function **f dtransform** to realize this conversion. Display the magnitude and log magnitude responses using **g dtft** and sketch the plots.

(b) Following the same general procedure, convert $H(z)$ to a bandpass filter with $\omega_{c1} = 0.4\pi = 1.2566$ and $\omega_{c2} = 0.5\pi = 1.5708$. Display the magnitude and log magnitude responses using **g dtft** and sketch the plots.

(c) Convert $H(z)$ to a bandstop filter with $\omega_{c1} = 0.3\pi = 0.9425$ and $\omega_{c2} = 0.6\pi = 1.885$. Display the magnitude and log magnitude responses using **g dtft** and sketch the plots.   □

EXERCISE 3.4.2.   A Digital Elliptic Bandpass Filter

Use **f aelliptic**, **f bilinear**, and **f dtransform** to design an elliptic digital bandpass filter of minimum order with $\omega_{s1} = 0.1\pi = 0.314159$, $\omega_{p1} = 0.2\pi = 0.6283$, $\omega_{p2} = 0.3\pi = 0.9425$, and $\omega_{s2} = 0.4\pi = 1.2566$. Note that ω_{s1} and ω_{p1} are the stopband and passband cutoff frequencies defining the first transition band, and ω_{p2} and ω_{s2} are the cutoff frequencies defining the second one. The ripples in the passband and in both of the stopbands should be 0.01. Use **g dfilspec** to examine the upper and lower cutoff frequencies to verify that the digital filter satisfies the specifications.

Although they are not laid out for you, this problem will require several steps. Since the digital passband and stopband cutoff frequencies are specified, the transition width of the analog filter must be estimated before the bilinear transform is taken and before frequency transformation. You should carefully record all of these steps in your write-up. What is the minimum order you were able to achieve given the specifications above? Display and sketch the pole/zero and magnitude response plots, using **g polezero**.   □

EXERCISE 3.4.3.   Digital Highpass Design

We have seen two ways to design highpass, bandpass, and bandstop digital filters from classical analog lowpass prototypes. The first is to perform a frequency band transformation on the analog filter and then to convert it to a digital filter using, for example, the bilinear transformation. The second is to convert the lowpass filter to a digital one and then perform the frequency band transformation. Here we compare these approaches for the design of a digital highpass filter with a cutoff frequency of $\omega_c = 0.3\pi = 0.9425$.

(a) Design a fifth-order analog Chebyshev I lowpass filter, $H_a(s)$, with $\epsilon = 0.33$. Use **g afreqres** with $\Omega_L = 0$ and $\Omega_H = 3$ to display the magnitude response and sketch the plot. The goal is to design a digital highpass filter with cutoff frequency 0.3π using the bilinear transformation. Apply **f atransform** to perform the highpass frequency transformation $s \rightarrow \Omega_d/s$, where Ω_d is a frequency you are to determine. Next apply **f bilinear** with $\beta = 2$ to convert the filter to digital form. Display its magnitude response using **g dtft** and sketch the plot.

(b) Design the same filter by applying the bilinear transform to $H_a(s)$ and then performing a digital frequency band transformation using **f dtransform**. The parameters of the transformation should be determined from the digital transformation expressions given in this subsection and should result in a highpass digital filter with nominal cutoff frequency of $0.3\pi = 0.9425$. Record the transformation expression that you used. Display the magnitude response using **g dtft** and sketch the plot. Record your value for Ω_d.

(c) How do these two approaches compare in terms of digital filter quality and ease of design? □

EXERCISE 3.4.4. Power Complementary Digital Filters

In Exercise 3.2.4 we said that two analog filters with frequency responses $G_a(j\Omega)$ and $H_a(j\Omega)$ are power complementary if their analog frequency responses satisfy the relation

$$|G_a(j\Omega)|^2 + |H_a(j\Omega)|^2 = 1.$$

Similarly, two digital filters are power complementary if their digital frequency responses satisfy the similar relation

$$|G(e^{j\omega})|^2 + |H(e^{j\omega})|^2 = 1.$$

(a) Show that, if $G_a(j\Omega)$ and $H_a(j\Omega)$ are power complementary analog filters, then the digital filters that result from them using the bilinear transformation are also power complementary for any value of the parameter β.

(b) Design a power complementary digital Butterworth lowpass/highpass filter pair of sixth order with 3-dB cutoff frequencies at $\omega_c = \pi/2$ using the functions **f abutter**, **f bilinear**, and **f dtransform**. Display the magnitude response of $G(z)$ using **g dtft**. Upon exiting the function, a file called _dtft_ is created that contains the DTFT values that were plotted to the screen. Copy this to the file **temp1**. Now display the magnitude response of $H(z)$ using **g dtft** and copy _dtft_ into the file **temp2**. You may use the **g look** function to display the frequency responses in **temp1** and **temp2** to verify that they are correct.

(c) Next compute the squared magnitude (or spectral power) $|G(e^{j\omega})|^2$ and $|H(e^{j\omega})|^2$ by typing

> **f multiply   temp1   temp1   temp3**
> **f mag   temp3   temp3**

and

> **f multiply   temp2   temp2   temp4**
> **f mag   temp4   temp4**

Using the **g slook2** function, display these squared magnitude responses superimposed on the same plot and sketch them. Now type

> **f add   temp3   temp4   temp5**

This will add the squared responses together. Use the **g look** function to display **temp5**. If the digital power complementary filters were designed properly, the displayed result should be a straight line with a constant amplitude of one. □

EXERCISE 3.4.5.   **Recursive Comb Filters**

A comb filter has a periodic frequency response. In the pole/zero domain, this corresponds to a radially symmetric pattern of zeros around the unit circle.

(a) Use the *change pole/zero* option in **g polezero** to create a filter by first adding a zero at $z = 1$. Return to the main menu and then add a pole at $z = 0.7$. Sketch the magnitude response and return to the main menu. Write the filter to a file called **hz**. Then exit the **g polezero** function.

(b) Now repeat the procedure in part (a) but this time add six zeros at $z = 0.5 \pm 0.866$, $z = -0.5 \pm 0.866$, and $z = \pm 1$. Then add six poles at $z = 0.471 \pm 0.816$, $z = -0.471 \pm 0.816$, and $z = \pm 0.9423$ and sketch the magnitude plot. Return to the main menu, write the filter to a file called **hz1**, and exit the program. Examine the filter **hz1** using **g dfilspec** and determine the frequencies at which the magnitude response goes to zero.

(c) Type the coefficients of **hz** and **hz1** to the screen and record the coefficient values. Note that values that are very small (less than 10^{-3}) can be viewed as zero. The fact that these values are not precisely zero is due to quantization and roundoff errors. What simple substitution allows you to transform **hz** into **hz1**. This simple substitution allows for the easy construction of comb filters.   □

3.5   A COMPLETE DESIGN PROCEDURE

The digital IIR filter design task begins with a set of digital filter specifications and then finds a digital filter that satisfies those conditions. The approach considered in this text is to first design an analog prototype and convert it to digital form. Most of the issues related to this design procedure have been addressed in the previous sections. The design of classical analog prototype filters was discussed and formulas were introduced for the Butterworth, Bessel, Chebyshev I, Chebyshev II, and elliptic filters. Each of these filters has properties that make it special. The elliptic filter is particularly attractive because it has optimal equiripple magnitude response characteristics.

Mapping procedures for converting analog prototypes into digital filters were also discussed and included impulse invariance, step invariance, the bilinear transform, the matched z-transform, and several other transformations. The bilinear transform is most often the method of choice.

Next, frequency transformations were introduced and formulas were given that allow lowpass prototypes to be converted into highpass, bandpass, bandstop, and other lowpass filters. Formulas were given for both analog domain and digital domain transformations. Either can be used, as was shown in the preceding exercises. However, digital transformations will be used in the complete design procedure outlined in this subsection. In addition to transformations, we discussed control of the prototype filter's cutoff frequency by using the T and β parameters in the impulse invariance and bilinear transform techniques, respectively.

The remaining issue, which has not been formally addressed, is the conversion from analog frequency specifications (in terms of Δ_p and Δ_s) to digital specifications (in terms of δ_p and δ_s). As stated in the previous chapters, the *digital* frequency specifications are typically given in terms of a passband deviation about unity. In other words, the deviation can be either positive or negative in the passband. The analog convention, however, restricts the maximum filter response value to be unity and the deviation, therefore, to be only negative. To convert an analog filter, $H_a(j\Omega)$, with a 1 to $1 - \Delta_p$ passband magnitude deviation specification into a digital filter, $H(e^{j\omega})$, with a $1 + \delta_p$ to $1 - \delta_p$ specifications, the following simple relationships are worth noting: $\Delta_p = 2\delta_p/(1 + \delta_p)$ and $\Delta_s = \delta_s/(1 + \delta_p)$. These relationships between (Δ_p, Δ_s) and (δ_p, δ_s) will allow you to specify the analog deviations directly from the digital specifications. Once $H_a(j\Omega)$ is designed and converted to digital form, it can be multiplied by the gain factor $1 + \delta_p$ so that the resulting digital filter will satisfy the specified digital tolerances.

At this point, we can consider a complete design procedure based on using the bilinear transformation where we start with digital specifications and design digital filters. Listed below are the basic steps for the design of a digital elliptic lowpass filter:

1. Convert the digital passband and stopband deviations into analog prototype specifications using the relationships

$$\Delta_p = 2\delta_p/(1 + \delta_p)$$

and

$$\Delta_s = \delta_s/(1 + \delta_p).$$

2. Convert the digital cutoff frequencies ω_p and ω_s into the analog cutoffs Ω_p and Ω_s. Recall that in the bilinear transform, these frequencies are related by

$$\Omega_p = 2\tan\frac{\omega_p}{2}$$

and

$$\Omega_s = 2\tan\frac{\omega_s}{2}$$

when β is equal to 2.

3. Compute the normalized analog cutoff frequencies $\hat{\Omega}_s$ and $\hat{\Omega}_p$, where

$$\hat{\Omega}_s = \Omega_s/\Omega_p$$

and

$$\hat{\Omega}_p = 1.$$

These are the frequencies that should be used in the analog design functions.

4. Design the analog elliptic lowpass prototype using **f aelliptic**.
5. Apply the bilinear transformation with

$$\beta = 2/\Omega_p.$$

This value of β counteracts the effect of using the normalized cutoff frequencies in the analog filter design stage.

6. Multiply the numerator of the filter by the gain factor $1 + \delta_p$ so that the resulting filter will have the proper passband amplitude range.

Digital Butterworth and Chebyshev filters can be designed in a similar fashion, as illustrated in the following exercises. In the event that a highpass, bandpass, or bandstop filter is desired, a digital lowpass prototype can be determined first. After its design, an appropriate digital frequency transformation can be applied.

Clearly this procedure is not the only one that can be followed. The order of operations can be arranged in many different ways. In the exercises that follow, you are asked to do complete designs of digital filters given digital frequency specifications. You are instructed to do these designs by performing several discrete steps and by using the functions provided in the software. This is done strictly for educational purposes. For practical applications where filters must be designed routinely, you would naturally combine these operations to create an automated design program.

EXERCISE 3.5.1. A Simple Digital Elliptic Lowpass Filter

In this exercise, you go through the steps outlined in this subsection to design a digital elliptic lowpass filter. You should design a filter of lowest order that satisfies the following constraints: $\omega_p = 0.993$ rad, $\omega_s = 1.2$ rad, $\delta_p \leq 0.1$, and $\delta_s \leq 0.026$.

(a) We begin by computing the analog prototype specifications

$$\Delta_p = 2\delta_p/(1 + \delta_p) = 0.181818$$

and

$$\Delta_s = \delta_s/(1 + \delta_p) = 0.02636.$$

Next we compute the analog cutoffs

$$\Omega_p = 2\tan\frac{\omega_p}{2} = 1.083533$$

and

$$\Omega_s = 2\tan\frac{\omega_s}{2} = 1.36827.$$

Normalizing these frequencies, we obtain

$$\hat{\Omega}_s = \Omega_s/\Omega_p = 1.2628$$

and

$$\hat{\Omega}_p = 1.$$

Use $\hat{\Omega}_s$, Δ_p, and Δ_s in the analog design function **f aelliptic** to design $H_a(j\Omega)$. Use **g afreqres** with $\Omega_L = 0$ and $\Omega_H = 3$ to display the magnitude response and sketch the plot.

(b) Apply the bilinear transformation using **f bilinear** with

$$\beta = 2/\Omega_p = 1.84.$$

Use **g dtft** to display the magnitude response of the digital filter and sketch it. You should observe that the maximum amplitude of the filter is one and not $1 + \delta_p$.

(c) Use the **f gain** function to multiply the numerator by the gain factor $1 + \delta_p = 1.1$. Display the magnitude response of the filter using **g dfilspec** and sketch the plot. Use the *right* and *left* arrow keys to record the passband and stopband ripples, and the passband and stopband cutoff frequencies. Verify that they are within the design tolerances.   □

EXERCISE 3.5.2.   Digital Chebyshev II Lowpass Filter

Following the procedure outlined in this subsection, design a digital Chebyshev II lowpass filter with the following magnitude specifications: $\omega_p = 0.5\pi = 1.57$ rad, $\omega_s = 2.18$ rad, $\delta_p \leq 0.05$, and $\delta_s \leq 0.01$. Notice that a modification to the design procedure given for elliptic filters is necessary when using the analog Chebyshev II filter design function. Recall that **f acheby2** requires a normalized stopband cutoff frequency (i.e., $\hat{\Omega}_s = 1$) and a passband cutoff frequency that is less than 1.

(a) Determine the normalized cutoff frequency, $\hat{\Omega}_p$, and the value of β for the bilinear transform function **f bilinear**.

(b) Design the digital filter. This will use the functions **f ocheby2** to determine the analog input parameters and **f acheby2** to design the analog prototype. Use **g dfilspec** to display the magnitude response of the digital filter and sketch the plot. Using the *right* and *left* arrow keys, determine the actual passband and stopband deviations at the cutoff frequencies. What is the minimum order digital Chebyshev II filter required to meet these specifications?   □

EXERCISE 3.5.3.   Digital Butterworth Lowpass Filter

Design a digital Butterworth lowpass filter with the following magnitude specifications: $\omega_p = 0.5\pi = 1.57$ rad, $\omega_s = 2.18$ rad, $\delta_p \leq 0.05$, and $\delta_s \leq 0.01$. Notice that a modification to the design procedure given for elliptic filters is necessary when using the analog Butterworth filter design function. Recall that **f abutter** requires a 3-dB cutoff as opposed to passband and stopband cutoff frequencies. Consequently, you must convert from Ω_p to the 3-dB frequency and use a filter designed with a normalized 3-dB frequency.

(a) Determine the value of β for the bilinear transform function **f bilinear**.

(b) Design the digital Butterworth filter. Use **g dfilspec** to display the magnitude response of the digital filter and sketch the plot. Using the *right* and *left* arrow keys, determine the actual passband and stopband deviations at the specified cutoff frequencies. What is the order of your filter?   □

EXERCISE 3.5.4.   Digital Butterworth Bandstop Filter

The design of a digital bandstop filter is an extension of the digital lowpass filter design problem. Consider the design of a digital elliptic bandstop filter with the following magnitude specifications:

- In the first passband region, the passband cutoff frequency is $0.2\pi = 0.628$ rad and the maximum passband deviation is 0.1.

- In the bandstop region, the upper and lower cutoff frequencies are $0.4\pi = 1.257$ and $0.7\pi = 2.199$ rad, respectively, and the maximum stopband deviation is 0.05.

- In the second passband region, the passband cutoff frequency is $0.8\pi = 2.513$ and the maximum passband deviation is 0.05.

First determine the design specifications for the digital lowpass prototype that will be used in the digital frequency transformation. Record these specifications and determine the specifications for the analog prototype. Next, design the digital filter. Display and sketch the magnitude response plot using **g dtft**. Then apply **f dtransform** to convert it to a digital bandstop filter. Display and sketch the magnitude response plot using **g dfilspec**. Using the *right* and *left* arrow keys, determine the actual maximum passband and stopband deviations for the given cutoff frequencies. What is the minimum filter order for the bandstop filter?   □

EXERCISE 3.5.5.   **Digital Chebyshev I Highpass Filter**

Design a digital Chebyshev I highpass filter with the following magnitude specifications: $\omega_s = 0.5\pi = 1.571$ rad, $\omega_p = 0.65\pi = 2.042$ rad, $\delta_s \leq 0.01$, and $\delta_p \leq 0.05$. At this point, you should be familiar with the steps involved in the procedure and the functions available to carry out the various steps. Record the values for Δ_p, Δ_s, and Ω_s used in the analog prototype, the value for β used in the bilinear transformation, and the parameters used for the digital frequency transformation. What is the order of your digital highpass filter? Use **g dfilspec** to display the magnitude response of the digital filter and sketch the plot. Using the *right* and *left* arrow keys, determine the actual passband and stopband deviations at the specified cutoff frequencies.   □

EXERCISE 3.5.6.   **A Fifth-Order Digital Chebyshev I Lowpass Filter**

Design a fifth-order digital Chebyshev I lowpass filter with a maximum passband deviation of 0.05 and maximum stopband deviation of 0.02 and passband cutoff frequency of $0.5\pi = 1.571$ rad. Using **g dfilspec**, display and sketch the magnitude response of the digital filter. Using the *right* and *left* arrow keys, determine the width of the transition band region.   □

EXERCISE 3.5.7.   **Comparing IIR and FIR filters**

In this problem, you will design several digital lowpass filters to meet the following specifications:

     (i) $\omega_p = 0.2\pi = 0.628$: passband cutoff frequency.

    (ii) $\omega_s = 0.3\pi = 0.942$: stopband cutoff frequency.

    (iii) $\delta_p = 0.005$; $\delta_s = .01$.

(a) Using **f obutter**, determine the minimum order digital Butterworth filter using the bilinear transformation that will meet these requirements. Note that you must first determine the lowpass analog prototype with a unity gain and a 3-dB cutoff frequency.

(b) Repeat for a Chebyshev I filter using **f ocheby1**.

(c) Repeat for a Chebyshev II filter using **f ocheby2**.

(d) Repeat for an elliptic filter using **f oelliptic**.

(e) Determine the minimum order linear phase Kaiser window filter that will meet the specifications above using **f eformulas** (option 1).

(f) Determine the minimum order linear phase equiripple filter that will meet the specifications above using **f eformulas** (option 3).

(g) Assuming that each filter is implemented with a single difference equation, determine the minimum number of multiplies and adds required to implement each filter.   □

REFERENCES

[1] A. V. Oppenheim and R. W. Schafer, *Discrete-Time Signal Processing,* Prentice-Hall: Englewood Cliffs, NJ, 1989.

[2] L. R. Rabiner and B. Gold, *Theory and Application of Digital Signal Processing,* Prentice-Hall: Englewood Cliffs, NJ, 1975.

[3] A. Antoniou, *Digital Filters: Analysis and Design*, McGraw-Hill: New York, 1979.

[4] T. W. Parks and C. S. Burrus, *Digital Filter Design,* John Wiley: New York, 1987.

Allpass and Multirate Filters 4

FIR and IIR filters are most often designed for implementation at fixed sampling rates, i.e., the filters are implemented using series of linear constant coefficient difference equations, each of which has the same time index n. Under initial rest conditions these filters are linear and time invariant. Filters, however, may also be designed for simultaneous operation over several sampling rates. These are called *multirate filters,* and their implementation includes upsamplers and downsamplers, which increase or decrease the sampling rate by an integer factor. Filters applied before and after these operators function at different rates and often have the property that their magnitude responses are approximately constant. When the magnitude responses are exactly constant, for $-\pi \leq \omega < \pi$, they are called *allpass* filters. Filters with allpass and approximate allpass characteristics are important building blocks in multirate implementations.

In the next section, allpass filters, their properties, and their application to problems are presented. *Polyphase filters* are then introduced, and their relationship to allpass filters is discussed. It is shown how multirate filters implemented using polyphase structures are sometimes able to achieve tremendous savings in computation for a given set of filter specifications. Additional exposure to this topic is given in Chapter 6, where several projects involving the design of multirate systems may be found.

Multirate filters and systems are often used to implement analysis/synthesis systems for subband speech and image coding and progressive digital transmission of images and video. Although the intention here is to only cover the basics, a very brief discussion on multirate analysis/synthesis subband systems is included. The reader interested in these and other aspects of multirate systems is referred to [6], [7], and [8], which provide a more complete and well-balanced discussion of multirate theory and its applications.

4.1 ALLPASS FILTERS

Allpass filters form an interesting class of systems that have been used for group delay equalization, multirate frequency selective filter implementation, notch filtering, transmultiplexing, frequency band transformations, and subband coding. They have a constant magnitude response at all frequencies. Thus, if $\mathcal{A}(e^{j\omega})$ is the frequency response of an arbitrary allpass filter, then

$$|\mathcal{A}(e^{j\omega})| = C, \qquad \text{for all } \omega.$$

The phase responses, however, will differ from one allpass filter to the next. A delay is a trivial example of an allpass filter. More general ones have system functions that are expressible in the form

$$\mathcal{A}(z) = C\frac{z^{-N}A^*(1/z^*)}{A(z)} = C\frac{a_0^*z^{-N} + a_1^*z^{-N+1} + \cdots + a_N^*}{a_0 + a_1 z^{-1} + \cdots + a_N z^{-N}} \tag{4.1}$$

where the asterisk "*" denotes the complex conjugate. The numerator and denominator polynomials of the allpass system function are related to each other in a very special way. We say that they form a conjugate mirror-image pair as indicated in (4.1). Commonly, the filter coefficients are real, in which case $a_i = a_i^*$, and the numerator is simply a shifted, time-reversed copy of the denominator. It is possible for an allpass filter to assume a form without the mirror-image or time-reversed relationship shown in (4.1) when the numerator and denominator share redundant roots. In such a case, when the redundant poles and zeros are removed the filter reduces to the form of (4.1).

Allpass filters can be used as components to implement digital filters efficiently. Due to the mirror image relationship of the filter coeffficients, an allpass filter, $\mathcal{A}(z)$, with real coefficients can be realized using the difference equation

$$
\begin{aligned}
y[n] \quad = \quad & a_1(x[n - (N - 1)] - y[n - 1]) + a_2(x[n - (N - 2)] - y[n - 2]) \\
& + \ldots + a_N(x[n] - y[n - N]) + x[n - N]
\end{aligned} \tag{4.2}
$$

where $x[n]$ is the input sequence, $y[n]$ is the output sequence, and $a_1, a_2, \ldots, a_N$ are the allpass filter coefficients. A difference equation of this form requires approximately half the multiplications of a direct form implementation. Moreover, the constant magnitude characteristic is preserved under coefficient quantization. This point and others related to the numerical properties of allpass filters in finite precision arithmetic environments are further discussed in Chapter 5.

Allpass filters can be used as efficient building blocks for implementing many classical IIR frequency selective filters. It is not the intent of this discussion to develop the theory for allpass decompositions as this would take us too far afield. Rather, our aim is to highlight the important properties and show how they can be used to decompose many common filters into the sum of two allpass filters for efficient implementation.

Consider an arbitrary Nth-order filter of the form

$$H(z) = \frac{B(z)}{A(z)} = \frac{\displaystyle\sum_{n=0}^{N} b[n]z^{-n}}{1 + \displaystyle\sum_{n=1}^{N} a[n]z^{-n}},$$

where both the numerator polynomial, $B(z)$, and the denominator polynomial, $A(z)$, are of degree N. This system function can be decomposed into the form

$$H(z) = \frac{1}{2}[\mathcal{A}_0(z) + \mathcal{A}_1(z)], \tag{4.3}$$

where $\mathcal{A}_0(z)$ and $\mathcal{A}_1(z)$ are allpass filters, if the following conditions hold:

1. $H(z)$ is stable, and $|H(z)| \leq 1$ for all ω. If this condition is valid, the filter is said to be *bounded real*.

2. The numerator polynomial $B(z)$ is a symmetric polynomial, i.e.,

$$b[n] = b[N - n]. \tag{4.4}$$

3. The spectral factor $D(z) = d_0 + d_1 z^{-1} + \ldots + d_N z^{-N}$ must be antisymmetric, i.e.,

$$d[n] = -d[N - n], \tag{4.5}$$

where

$$D(z)D^*(1/z^*) = A(z)A^*(1/z^*) - B(z)B^*(1/z^*). \tag{4.6}$$

Although these are not necessary conditions, neither are they particularly stringent and they do provide a convenient mechanism for performing the decomposition.

Filters meeting the above-stated conditions may be decomposed into the form of (4.3) in the following way.

Procedure 1

The first step is to determine $D(z)$, the antisymmetric spectral factor in (4.6), by factoring the polynomial $A(z)A^*(1/z^*) - B(z)B^*(1/z^*)$. Once $D(z)$ is determined, the allpass filters may be obtained by computing the roots of the Nth-order polynomial $B(z) + D(z)$. These roots should be separated into two groups: the first consisting of the M roots inside the unit circle $\theta_1, \theta_2, \ldots, \theta_M$, which are used to form $\mathcal{A}_0(z)$, and the second consisting of the L roots outside the unit circle $\gamma_1, \gamma_2, \ldots, \gamma_L$, which are used to obtain $\mathcal{A}_1(z)$.

The allpass filters $\mathcal{A}_0(z)$ and $\mathcal{A}_1(z)$ are then defined as

$$\mathcal{A}_0(z) = \prod_{i=1}^{M} \frac{\theta_i^* - z^{-1}}{1 - \theta_i z^{-1}} \tag{4.7}$$

and

$$\mathcal{A}_1(z) = \prod_{i=1}^{L} \frac{(1/\gamma_i^*) - z^{-1}}{1 - (1/\gamma_i)z^{-1}}, \tag{4.8}$$

where $M + L = N$. A discussion of the theory leading to this procedure can be found in [2] and early development is found in [3].

Of greatest interest to us is the allpass decomposition of classical Butterworth, Chebyshev, and elliptic filters. As it turns out, the three conditions listed above are satisfied by these classical IIR filters when the filter order is odd and the filter is a lowpass or highpass. (These conditions are satisfied for bandpass and bandstop filters when the order is two times an odd integer.) These classical odd-order filters display several important properties, which lead to a simplification of the above procedure. First, each pole in $H(z)$ is forced to appear in either $\mathcal{A}_0(z)$ or $\mathcal{A}_1(z)$. This follows directly from (4.3). Second, assume a locus of points is drawn connecting nearest neighbor poles of $H(z)$ in the first and second quadrants of the z-plane and that these poles are numbered in order of increasing position on the locus. Then these poles will typically appear in pairs alternately in $\mathcal{A}_0(z)$ and $\mathcal{A}_1(z)$. In other words, the first, third, fifth, etc., pole of $H(z)$ will go to one allpass filter, and the second, fourth, sixth, etc., will go to the other. Since the filter is constrained to be odd order, complex conjugate poles will not be split between $\mathcal{A}_0(z)$ and $\mathcal{A}_1(z)$, and consequently, real allpass filters will result. This property is derived in [13] for the classical analog elliptic filter and is generalizable to the Butterworth and Chebyshev filters. Since the bilinear transformation will not disturb the ordering of the poles, the property is valid in the discrete-time domain as well. Finally, the poles of an allpass filter completely specify the filter because the numerator and denominator are reversed or mirror imaged versions of each other. Thus, this simple alternating distribution property of the poles of classical IIR filters provides a simple way to perform the allpass decomposition. This is summarized in the procedure below.

Procedure 2

1. Factor the denominator of $H(z)$.
2. Distribute the poles between $\mathcal{A}_0(z)$ and $\mathcal{A}_1(z)$ based on the alternating selection process just described.
3. Convert the poles into polynomials $D_0(z)$ and $D_1(z)$.
4. Construct the allpass filters $\mathcal{A}_0(z)$ and $\mathcal{A}_1(z)$, using the fact that the numerator and denominator polynomials are time reversed copies of each other. When doing this step using the software, make sure that both filters have starting points at zero.

Allpass decomposition of IIR filters is further illustrated and developed in the exercises that follow. Allpass filters are also useful when cascaded with other filters. Although all of these filters have constant magnitude responses, each may be designed to have a prescribed group delay response. Consequently, they are ideal for group delay compensation, since when they are cascaded with other filters they will alter the overall group delay response without affecting the magnitude response. In the remainder of this section, allpass filters will be examined with respect to their special properties and in the context of some of their applications.

EXERCISE 4.1.1. **Pole/Zero Relationship for Allpass Filters**

The number of zeros in an allpass system function is equal to the number of its poles. Furthermore, the zeros are the complex conjugate reciprocals of the poles. This exercise explores this relationship.

(a) Use the **g polezero** function to create a filter with a pole at $z = 0.8$ and a zero at $z = 1.2$. This may be done in two steps using the *change pole/zero* option. First add a pole to the z-plane, then return to the main menu and repeat the procedure to add the zero.

Sketch the magnitude response of the filter. Now move the zero to $z = 1.25$ and update the magnitude response that is displayed on the screen by pressing the "d" key as you move the zero. Sketch the magnitude response when the pole is at $z = 0.8$ and the zero is at $z = 1.25$. Observe the effect of perturbing this pole/zero pair on the magnitude response. Write this allpass filter to a file **temp** using the *write to output file* option and then exit the program.

(b) Following the same procedure, create another filter with a complex pole at $z = j(0.8)$ and a complex zero at $z = j(1.2)$. You may wish to use the *adjust circle size* option to rescale the plot so that both the pole and the zero are shown in the field of view.

Sketch the magnitude response. Use the *change pole/zero* option to move the zero to $z = j(1.25)$ and sketch the magnitude response. Save this filter in the file **temp1**.

(c) Use the **f cas** function to cascade the allpass filters in **temp** and **temp1** together. Use **g dtft** to display and sketch the magnitude response of the result.

These two filters have the form

$$H(z) = \frac{z^{-1} - a^*}{1 - az^{-1}}. \tag{4.9}$$

Notice that the pole and zero occur at conjugate reciprocal positions. Using the substitution $z = e^{j\omega}$, show analytically that the magnitude response of $H(e^{j\omega})$ is one for all ω. Also, show that a cascade of systems of the form of (4.9) will also be an allpass. Show that any such cascade will also have the form shown in (4.1).   □

EXERCISE 4.1.2.   Allpass Filter Decomposition

Filters that are bounded real and satisfy the symmetry conditions given in (4.4) and (4.5) can be decomposed into sums of allpass filters. In this exercise, we apply the allpass decomposition procedure discussed in this section introduction to the filter $H(z)$, where

$$H(z) = \frac{B(z)}{A(z)} = \frac{0.045 + 0.1375z^{-1} + 0.1375z^{-2} + 0.045z^{-3}}{1 - 1.36z^{-1} + 1.02z^{-2} - 0.295z^{-3}}. \tag{4.10}$$

(a) Use the *create file* option in **f siggen** to create a file **hz** for $H(z)$. Using **g dtft**, sketch its magnitude response and verify that it is a bounded real function.

Next construct the polynomial $A(z)A^*(1/z^*) - B(z)B^*(1/z^*)$. Notice that the coefficients of $A(z)$ and $B(z)$ are real and therefore have either real roots or roots in complex conjugate pairs in the z-plane. Consequently, $A^*(1/z^*)$ and $B^*(1/z^*)$ are equivalent to $A(1/z)$ and $B(1/z)$ and correspond to simple reversals of the $A(z)$ and $B(z)$ polynomials.

The polynomial $A(z)A^*(1/z^*) - B(z)B^*(1/z^*)$ can be computed conveniently by using the macro **qpoly.bat** which, for convenience, is provided for you on the disk. The macro may be invoked by typing

$$\textbf{qpoly}\quad\textbf{hz}\quad\textbf{output}$$

where **output** is the name of the file containing the polynomial.

(b) Display and sketch the magnitude and pole/zero plots of $A(z)A^*(1/z^*)-B(z)B^*(1/z^*)$ using **g polezero**. Identify the roots that correspond to the spectral factor $D(z)$. You will observe that the pole/zero plot contains six roots arranged in three double zero pairs. Obtain the spectral factor $D(z)$ by deleting one from each pair using the *change pole/zero* option and the *write to output file* option to write $D(z)$ to a file.

Next use **f add** to compute $D(z)+B(z)$. The **f convert** function can be used to obtain $B(z)$ from $H(z)$. Use the *change pole/zero* option in **g polezero** to delete all roots of $D(z)+B(z)$ outside the unit circle and write the result to a file; call this polynomial $F_0(z)$. Then read in the same file again but this time delete all roots inside the unit circle. Write this result to another file and call the result $F_1(z)$. The polynomials $F_0(z)$ and $F_1(z)$, although purely real, are output as complex numbers (with zero imaginary parts) by the **g polezero** function. Use **f realpart** to convert these polynomials into purely real representations.

The polynomials, $F_0(z)$ and $F_1(z)$, can now be used to form $\mathcal{A}_0(z)$ and $\mathcal{A}_1(z)$ using **f reverse**, and **f lshift** to form the mirror image polynomial and **f revert** to form the IIR allpass filter. Construct $\mathcal{A}_0(z)$ and $\mathcal{A}_1(z)$ and display and sketch the group delay responses using **g dtft**. Examine these filters using **g polezero** and verify that both are causal and stable, i.e., the poles are inside the unit circle. Sketch the pole/zero plot.

(c) Add $\mathcal{A}_0(z)$ and $\mathcal{A}_1(z)$ together using the **f par** function to form a lowpass filter. Examine the result using **g dtft** and sketch the magnitude response. Observe that $\mathcal{A}_0(z)+\mathcal{A}_1(z)=2H(z)$. $\qquad\square$

EXERCISE 4.1.3. **Complementary Filters**

Two of the exercises in Chapter 3 looked at power complementary filters. However, the concept of complementary filters is more general as we shall see. Complementary filters have the general property that they sum to unity in some way or another and have some interesting connections with allpass filters. There are several types of complementary filters that shall be considered in this exercise.

1. The filters $H_0(e^{j\omega})$ and $H_1(e^{j\omega})$ are said to be *direct-sum complementary* if they satisfy the relationship

$$H_0(e^{j\omega})+H_1(e^{j\omega})=1.$$

Filters with this property are most often just called *complementary*. However, to avoid possible confusion with the other types of complementary filters, we will refer to $H_0(e^{j\omega})$ and $H_1(e^{j\omega})$ as being *direct-sum* complementary. These filters are important in applications where several output channels from an analysis filter bank must be directly summed together, as is the case for audio systems with multiple loudspeakers.

2. The filters $H_0(e^{j\omega})$ and $H_1(e^{j\omega})$ are said to be *allpass complementary* if

$$|H_0(e^{j\omega})+H_1(e^{j\omega})|=1.$$

In other words, the sum of $H_0(e^{j\omega})$ and $H_1(e^{j\omega})$ should be allpass, which results in no spectral magnitude distortion, only changes in the spectral phase. Such filters have been suggested for certain speech applications where some level of phase distortion can be tolerated.

3. A pair of *power complementary* filters, as discussed in Chapter 3, satisfies the relationship

$$|H_0(e^{j\omega})|^2 + |H_1(e^{j\omega})|^2 = 1.$$

Power complementary filters have applications in subband coding and other multirate analysis/synthesis systems.

4. The filters $H_0(e^{j\omega})$ and $H_1(e^{j\omega})$ are said to be *doubly complementary* if they are both allpass complementary and power complementary. Such filters have been shown to have very computationally efficient realizations since they can be expressed in terms of allpass filters. Efficient implementations of allpass filters are discussed later in this chapter.

It should be noted that although our definitions for direct-sum, allpass, power, and doubly complementary filters require the terms to sum to unity, they could have been defined more generally to sum to a constant, C. Complementary filters in which this constant $C = 1$ are sometimes said to be *unit-magnitude*.

(a) Design a length 15 Hamming window lowpass filter, $h_a[n]$, with $\omega_c = \pi/2 = 1.5708$ using the function **f fdesign**. Shift it by -7 using **f lshift** so that it will be a zero-phase filter. Next, construct $h_b[n]$, where $H_b(e^{j\omega}) = 1 - H_a(e^{j\omega})$ or equivalently $h_b[n] = \delta[n] - h_a[n]$. This can be done easily using **f subtract** together with an impulse designed using **f siggen**. Display and sketch the real part of the DTFT for both $h_a[n]$ and its complement $h_b[n]$ using **g dtft**. Notice that both DTFTs are purely real and will sum to unity as desired. You may wish to examine the imaginary parts and verify that they are zero.

(b) Filters that are decomposable into a sum of allpass filters have complementary properties. In the previous exercise, it was shown that if a filter, $H_0(e^{j\omega})$, was bounded real and satisfied the symmetry conditions in (4.4) and (4.5), then it could be written as a sum of allpass filters,

$$H_0(e^{j\omega}) = \frac{1}{2}[A_0(e^{j\omega}) + A_1(e^{j\omega})]. \tag{4.11}$$

Its allpass complement, $H_1(e^{j\omega})$, is given by

$$H_1(e^{j\omega}) = \frac{1}{2}[A_0(e^{j\omega}) - A_1(e^{j\omega})]. \tag{4.12}$$

Let

$$A_0(z) = \frac{-0.20356 + z^{-1}}{1 - 0.20356 z^{-1}}$$

and

$$A_1(z) = \frac{0.66715 - 0.18053 z^{-1} + z^{-2}}{1 - 0.18053 z^{-1} + 0.66715 z^{-2}}.$$

Write these allpass filters to files using the *create file* option in **f siggen**. Construct $H_0(z)$ and $H_1(z)$ using the **f par** function and display and sketch the magnitude and phase for both filters. Again use the **f par** function to add $H_0(z)$ and $H_1(z)$. In the case of $H_1(z)$, the **f gain** function can be used to multiply $\mathcal{A}_1(z)$ by -1 prior to addition. Display and sketch the magnitude and phase responses for both filters using **g dtft**. Observe, analytically, that the allpass complementary property must always hold if the filters are sums of allpass filters as indicated above.

(c) Given the filters $H_0(e^{j\omega})$ and $H_1(e^{j\omega})$ in (4.11) and (4.12), prove analytically that they are power complementary.

(d) Are all filters that are allpass complementary also power complementary? If not, find an exception.

(e) Is it possible for a pair of filters to be both direct-sum complementary and power complementary? Carefully justify your answer.

As a final comment with respect to complementary filters, it should be noted that this concept is not limited to filter pairs. The number of filters that can form a set of complementary filters is arbitrary in general. □

EXERCISE 4.1.4.   Allpass Decomposition of an Elliptic Filter

In this exercise, you consider a classical digital elliptic filter and apply the allpass decomposition based on the alternating pole distribution property. Begin by designing a fifth-order analog elliptic filter with stopband cutoff frequency 1.3 rad/s, passband ripple 0.01, and stopband ripple 0.03, using **f aelliptic**. Then apply **f bilinear** with $\beta = 2$ to form $H(z)$.

(a) Display and sketch the magnitude response of $H(z)$ using **g dtft**.

(b) Following Procedure 2, outlined in the section introduction, decompose $H(z)$ into two allpass filters $\mathcal{A}_0(z)$ and $\mathcal{A}_1(z)$. Use the *change pole/zero* option and *write to output file* option in **g polezero** to split the poles of $H(z)$ between two output files according to the alternating pole property. The function **f convert** can be used to separate the numerator and denominator from the IIR filter. The functions **f reverse**, **f lshift**, and **f revert** can be used to construct the allpass filters, $\mathcal{A}_0(z)$ and $\mathcal{A}_1(z)$, from the denominator polynomials.

Examine their magnitude responses using **g polezero** and verify that they are allpass. Display and sketch the group delay responses for both allpass filters using **g dtft**.

(c) Use the **f par** function to add or combine $\mathcal{A}_0(z)$ and $\mathcal{A}_1(z)$ into a direct form filter and display the magnitude response of the filter using **g dtft**. Sketch the plot and verify that the decomposition procedure worked correctly.

(d) The filter, $H(z)$, can be implemented in direct form using the difference equation

$$
\begin{aligned}
y[n] \quad = \quad & -a_1 y[n-1] - a_2 y[n-2] - \ldots - a_5 y[n-5] \\
& + b_0 x[n] + \ldots + b_5 x[n-5],
\end{aligned}
$$

where $x[n]$ is the input, $y[n]$ is the output, and $\{a_i, b_i\}$ are the filter coefficients. How many multiplications and adds are required if this equation is used to implement the filter?

(e) If $H(z)$ is now implemented as a sum of allpass sections using difference equations of the form of (4.2), how many multiplications and adds are required? This computational reduction is an important property of allpass filters. □

EXERCISE 4.1.5. Allpass Decomposition of a Chebyshev II Filter

This exercise considers the allpass decomposition for a classical digital Chebyshev II lowpass filter. Begin by first designing a fifth-order analog Chebyshev II filter with $\epsilon = 0.2$ using **f acheby2**. Next apply **f bilinear** with $\beta = 2$ to form $H(z)$.

(a) Display and sketch the magnitude response of $H(z)$ using **g dtft**.

(b) Decompose $H(z)$ into two allpass filters following Procedure 2 outlined in the section introduction. To do this, use the *change pole/zero* option and *write to output file* option in **g polezero** to split the poles of $H(z)$ between two output files according to the alternating pole property. The function **f convert** can be used to separate the numerator and denominator from the IIR filter. The functions **f reverse**, **f lshift**, and **f revert** can be used to construct the allpass filters $\mathcal{A}_0(z)$ and $\mathcal{A}_1(z)$ from the denominator polynomials.

  Construct $\mathcal{A}_0(z)$ and $\mathcal{A}_1(z)$ and display the group delay responses using **g dtft**. Sketch these plots. Verify that $\mathcal{A}_0(z) + \mathcal{A}_1(z)$ produces a lowpass filter by using the **f par** function to combine this parallel combination of filters into a single one. Plot the magnitude response using **g dtft**.

  Display and sketch the group delay responses for both allpass filters using **g dtft**.

(c) Now consider the power complementary filter

$$G(z) = \frac{1}{2}[\mathcal{A}_0(z) - \mathcal{A}_1(z)].$$

Use the **f gain** function to form $-\mathcal{A}_1(z)$ and the **f par** function to convert $G(z)$ into direct form. Use **g dtft** to display the magnitude response of $G(z)$ and sketch it. What kind of filter is it; i.e., is it a Butterworth, Chebyshev I, Chebyshev II, Bessel, etc.? Explain why. □

EXERCISE 4.1.6. Allpass Decomposition of a Butterworth Filter

This exercise mimics the previous two. Here, however, you are asked to decompose a digital Butterworth lowpass filter into the sum of two allpass filters. Design a third-order analog Butterworth filter using **f abutter** and apply the bilinear transformation with $\beta = 2$ to form $H(z)$.

(a) Display and sketch the magnitude response of $H(z)$ using **g dtft**.

(b) Following Procedure 2 outlined in the section introduction, decompose $H(z)$ into two allpass filters. Use the *change pole/zero* option and *write to output file* option in **g polezero** to split the poles of $H(z)$ between two output files according to the alternating pole property. The function **f convert** can be used to separate the numerator and denominator from the IIR filter. The functions **f reverse**, **f lshift**, and **f revert** can be used to construct the allpass filters, $\mathcal{A}_0(z)$ and $\mathcal{A}_1(z)$, from the denominator polynomials. Use **g dtft** to sketch the group delay responses for both allpass filters.   □

EXERCISE 4.1.7.   **Allpass Decomposition of a Bessel Filter**

The last few exercises investigated allpass decompositions for Butterworth, Chebyshev, and elliptic filters. Here we consider the allpass decomposition for a Bessel filter. Design a fifth-order analog Bessel filter using **f abessel** and apply the **f bilinear** function with $\beta = 2$ to form $H(z)$.

(a) Display and sketch the magnitude response of $H(z)$ using **g dtft**.

(b) Can this digital Bessel filter be decomposed into the sum of allpass filters? Explain why or why not.

(c) Now attempt to perform the allpass decomposition using Procedure 1 outlined in the section introduction. Follow the steps given in Exercise 4.1.2, which describe the software functions that should be employed to perform the decomposition. What problem arises when trying to implement this procedure?

(d) Procedure 2, which is based on the alternating pole distribution property, would not seem to suffer from the problems encountered in Procedure 1. Perform the allpass decomposition using Procedure 2. Use the description in the last exercise for guidance on how to use the software to implement the decomposition. Unlike Procedure 1, Procedure 2 produces two allpass filters. Use the **f par** function to combine the allpass filters into a direct form filter, $G(z)$. Display and sketch the magnitude response of $G(z)$ using **g dtft** and compare it to the response of $H(z)$. Observe that although Procedure 2 allows you to obtain an answer, the resulting decomposition is not equivalent to the original filter.

Digital Bessel filters are not among the classical filters that are decomposable by this procedure. Fortunately, this is of small consequence because the usefulness of digital Bessel filters is very limited.   □

EXERCISE 4.1.8.   **Group Delay Equalization**

Since allpass filters have constant magnitude, they do not affect the magnitude response when they are cascaded with another filter, but they will affect the overall phase response. Therefore, they can be used to compensate for phase or group delay distortions in a system. When used in this fashion, they are known as *phase compensators* or *group delay compensators*. In many applications, the goal of phase compensation is to achieve a nearly constant group delay in the passbands and transition bands of the overall system. There is rarely a need for phase compensation in the stopband.

(a) Consider a second-order allpass filter that has complex poles at $z = re^{\pm j\theta}$. Its system function is given by

$$\mathcal{A}(z) = \frac{r^2 - 2r\cos\theta z^{-1} + z^{-2}}{1 - 2r\cos\theta z^{-1} + r^2 z^{-2}}.$$

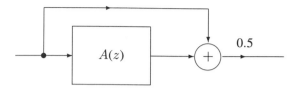

Figure 4.1. Implementation of a digital notch filter using an allpass.

Use the *create file* option of **f siggen** to create a second-order allpass filter for the case $\theta = \pi/2$, $r = 0.9$. Display and sketch the group delay of the filter using **g dtft**. Repeat for $\theta = \pi/4$, $r = 0.8$. You should observe that the group delay plot has a well-defined peak. How does the frequency at which the peak occurs depend on θ? How does the amplitude of the peak depend upon r?

(b) You should have observed in part (a) that the group delay of your allpass filters was positive for all frequencies. This is, in fact, a property that can be proven for all allpass filters. What this means is that we can only equalize the group delay by increasing it at frequencies where it is low, not by decreasing it at frequencies where it is high.

Design a second-order Butterworth lowpass filter using **f abutter**, and **f bilinear** with $\beta = 2$. Display the DTFT using **g dtft** and observe that the half-power cutoff frequency is about 0.93 rad. Display and sketch the group delay of the digital filter using **g dtft**. Find the largest and smallest values of the group delay over the interval $0 \le \omega \le \pi/4$. Record the difference between these values as a measure of the group delay deviation.

(c) In this part of the exercise, you should cascade the Butterworth filter that you designed in part (b) with a second-order allpass filter that is similar to the ones that you designed in part (a). Cascade the allpass filter with the Butterworth using **f cas**. You can type the macro to the screen to read the instructions for its use. Try to select the values of θ and r in the allpass to minimize the overall group delay deviation over the frequency range $0 \le \omega \le \pi/4$. Record your values for θ and r as well as the total deviation that you achieved.

(d) *(optional)* Repeat part (c) using two second-order allpass sections.   □

EXERCISE 4.1.9. Digital Notch Filter

A digital notch filter is useful for removing a signal component at a single frequency, such as 60 Hz interference or the carrier signal in certain communications systems. Regalia et al. [1] showed that a useful notch filter can be realized using an allpass filter in the configuration shown in Figure 4.1. The transfer function realized by this system is

$$H(z) = \frac{1}{2}[1 + \mathcal{A}(z)],$$

where

$$A(z) = \frac{r^2 - 2r\cos\theta z^{-1} + z^{-2}}{1 - 2r\cos\theta z^{-1} + r^2 z^{-2}}.$$

The notch frequency ϕ can be determined by the expression

$$\cos\phi = \frac{2r\cos\theta}{1 + r^2}.$$

(a) To implement the notch filter, first create an IIR filter with a zero-order numerator, first-order denominator, starting point 0, $b_0 = 1$, and $a_1 = 0$, using the *create file* option in **f siggen** and call the output file **one**. Next, create the second-order allpass filter, $A(z)$, with $\theta = \pi/2$ and $r = 0.8$. When added together, these filters will result in $2H(z)$. The function, **f par**, can be used to do this. Display and sketch the frequency response of $2H(z)$ using **g polezero**.

(b) Using a text editor or **f siggen**, vary the value of r in the range $0 < r < 1$ and examine the magnitude response of $2H(z)$ in each case. Determine an empirical formula that relates the width of the notch filter to the parameter r. For convenience, define the notch width to be the width at the 3dB frequencies of the filter.

(c) Examine the properties of $A(z)$ as discussed in the previous exercise and explain why θ controls the notch frequency of the filter and r controls the notch width. Why is r restricted to be between 0 and 1.

(d) Notch filters with multiple notch frequencies can be designed by cascading. Using the functions **f cas** and **f par**, design a filter with notch frequencies at $\pi/3 = 1.0472$ and $\pi/4 = 0.7854$, both with notch widths less than $\pi/8 = 0.3927$. Record your values of $theta$ and r used in each of the constituent filters. Display and sketch the magnitude response, using **g dtft**.   □

EXERCISE 4.1.10.   Tunable Filters

The system shown in Figure 4.2 is a tunable filter that can be used in a digital audio system to compensate for the frequency response in an acoustic playback environment. It can boost the amplitude of the audio signal in a region of the spectrum. Control over the shape of the magnitude response is provided by three parameters r, θ, and K.

The tunable filter consists of a weighted parallel combination of the digital notch filter from the previous exercise and its amplitude complementary filter. For the purpose of this exercise, let the allpass filter be the same second-order allpass that was used in the previous exercise. Here the parameters, r and θ, provide control over the bandwidth and center frequency just as they did for the notch filter.

(a) Determine the effect of the parameter K on the tunable filter. To do this, write a macro that implements the filter shown in Figure 4.2. Use the file **one** defined in Exercise 4.1.9a as the input to the structure in Figure 4.2. Your macro should contain the functions **f convert**, **f convolve**, **f add**, **f subtract**, **f gain**, and **f revert**. Write your macro so that it receives the allpass filter and K parameters as inputs and returns the tunable filter as an output.

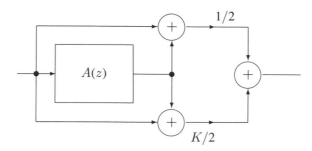

Figure 4.2. A tunable audio filter built from an allpass.

Use the macro to design several tunable filters with different values of K, all in the range $-1 < K < +1$. Describe the relationship between K and the shape of the tunable filter. Try to obtain an empirical formula that can be used for design purposes.

(b) Design a length 16 lowpass filter, $h[n]$, with cutoff frequency $0.4\pi = 1.256637$ using the *Hamming* window option in **f fdesign**. Construct the filter $v[n]$ defined as $v[n] = h[2n]$ and then upsample $v[n]$ by a factor of 2 to form $f[n]$. This can be done easily using the functions **f convolve**, **f dnsample**, and **f upsample**. The filter $f[n]$ will be used to represent an audio system with magnitude distortion. Display and sketch the magnitude response of $f[n]$ using **g dtft**.

Use your empirical formula and macro to design a tunable filter that can be cascaded with $f[n]$ to reduce the dip in the magnitude response. Your goal is to make the magnitude response of the result as flat as possible. Cascade the tunable filter with $f[n]$ using **f filter** with output length 120. Display and sketch the result using **g dtft**. Record the values of the parameters used in your tunable filter. ☐

4.2 MULTIRATE FILTERS

Filters that have components that are operating at more than one sampling rate are called *multirate filters*. Multirate implementations can often be very efficient. In addition to the components that we have already seen, multirate filters also contain downsamplers and upsamplers. These operators and their properties were discussed in the companion text[1] but are summarized here for convenience.

An M-to-1 *downsampler* (commonly denoted by a box labeled with $\downarrow M$) operates on successive blocks of M samples by retaining the first sample in each block and discarding the following $M - 1$ samples. If $x[n]$ is the input sequence and $y[n]$ is the output sequence, we can express the operation of a downsampler as

$$y[n] = x[Mn].$$

---

[1]M. J. T. Smith and R. M. Mersereau, *Introduction to Digital Signal Processing: A Computer Laboratory Textbook*, John Wiley: New York, 1992.

In the frequency domain, the Fourier transforms of $x[n]$ and $y[n]$ are related by

$$Y(e^{j\omega}) = \frac{1}{M} \sum_{r=0}^{M-1} X(e^{j(\frac{\omega}{M} + \frac{2\pi r}{M})}). \tag{4.13}$$

Downsampling does two important things to the signal spectrum. First, it introduces aliasing in the output. All spectral information in $x[n]$ at frequencies higher than π/M is folded into this band and added to what was previously there. Second, the frequency axis is stretched by a factor of M.

The *upsampler* is the dual operation. A 1-to-M upsampler (denoted by a box containing $\uparrow M$) inserts $M-1$ zeros between adjacent input samples. If $x[n]$ is the input to the upsampler and $y[n]$ is its output, the time signals are related by

$$y[n] = \sum_{m=-\infty}^{\infty} x[m]\delta[n - Mm],$$

and their spectra satisfy

$$Y(e^{j\omega}) = X(e^{j\omega M}).$$

In the z-domain this becomes $Y(z) = X(z^M)$. The important feature common to both operators is that they scale the frequency axes—expansion for downsampling and contraction for upsampling. The expansion and contraction of the spectrum are important for the implementation of efficient multistage multirate filters.

An M-to-1 *decimator* consists of a lowpass filter with a cutoff frequency of π/M followed by an M-to-1 downsampler. Its dual, a 1-to-M *interpolator*, is a cascade of a 1-to-M upsampler followed by a lowpass filter with a cutoff frequency π/M. If the cascade of an ideal interpolator and an ideal decimator is to act like an identity system, the interpolation filter must include a gain factor of M to compensate for the $1/M$ scale factor introduced into the spectrum by the M-to-1 decimator.

EXERCISE 4.2.1.   Downsampling and Upsampling a Filter

Downsamplers and upsamplers are basic components of multirate systems. This exercise will investigate the behavior of these operators.

(a) Design a length 64 FIR lowpass filter $h[n]$, using **f fdesign** with a Hamming window and a cutoff frequency of 0.5 rad. Display and sketch its magnitude response using **g dfilspec**. Then use **f dnsample** to downsample $h[n]$, first by a factor of two, then by a factor of three. Display and sketch the resulting magnitude responses. Carefully examine the resulting filters and describe the effect of downsampling on the filter's attenuation, center cutoff frequency, transition width, and filter length. Verify your observed results analytically.

(b) Upsample $H(z)$ using **f upsample**, first by a factor of two, then by a factor of three. Display and sketch the resulting magnitude responses using **g dtft**. Describe the effect of upsampling on the filter's attenuation, transition width, and filter length. Why did the lowpass filter become a multiband bandpass filter after upsampling? □

EXERCISE 4.2.2.   **Multirate Identities**

Filters can be moved from one side of an upsampler or downsampler to the other as long as the movement is from the lower rate side of the rate changer to the higher rate side. The result will be equivalent, if the appropriate modification is made to the system function of the filter. This point will be made clearer in this exercise.

To examine this property, use **f siggen** to design a signal, $x[n]$, consisting of four periods of a triangular wave with period 12. Use the signal in file **f001** as your filter $H(z)$.

(a) First use **f dnsample** to downsample $x[n]$ by a factor of seven. Use **f convolve** to filter this result with $H(z)$ and call the output $y_1[n]$. Next, try this in reverse order, using $H(z^7)$ as the filter. To be specific, generate $H(z^7)$ using **f upsample**. Filter $x[n]$ with $H(z^7)$ using **f convolve** and downsample the result by a factor of 7 using **f dnsample**. Call this result $y_2[n]$. Display and sketch $y_1[n]$ and $y_2[n]$ using **g view2**. Also examine these signals using **g sview2**. Are the results identical? Prove analytically that an M-to-1 downsampler followed by a filter $H(z)$ is equivalent to a filter $H(z^M)$ followed by an M-to-1 downsampler. Draw block diagrams of the two equivalent systems.

(b) Now filter $x[n]$ with $H(z)$ and upsample the result by a factor of 7. Call the result $y_3[n]$. Then upsample $x[n]$ by 7 and filter the result with $H(z^7)$ to form $y_4[n]$. Compare and sketch $y_3[n]$ and $y_4[n]$ as in part (a). Prove analytically that $H(z)$ followed by a 1-to-M upsampler is equivalent to a 1-to-M upsampler followed by $H(z^M)$. Again draw block diagrams of the two equivalent systems.   □

4.2.1 Polyphase Filters

Direct implementation of decimators and interpolators is computationally intensive, because the filtering that is inherent in these operations is performed at the higher sampling rate. One approach to improving computational efficiency is to modify the multirate structure so that the filtering is performed at the lower rate. *Polyphase structures,* which use *polyphase filters,* are based on this idea. The polyphase implementation for a decimator is shown in Figure 4.3. The polyphase filters, $p_0[n], p_1[n], \ldots, p_{M-1}[n]$ are related to the lowpass decimation filter, $h[n]$, by the expression

$$p_k[n] = h[nM + k]. \tag{4.14}$$

Notice that although there are now M polyphase filters instead of just one decimation filter, each polyphase filter contains only about $1/M$ as many coefficients. Moreover, all of the filtering is now performed at the lower sampling rate. We also notice from the figure that the decimation lowpass filter can be expressed as a parallel combination of delayed polyphase filters

$$H(z) = P_0(z^{-M}) + z^{-1}P_1(z^{-M}) + z^{-2}P_2(z^{-M}) + \ldots + z^{-(M-1)}P_{M-1}(z^{-M}).$$

Polyphase filters can be similarly exploited to implement interpolators using the structure in Figure 4.4. In this case, as before, the polyphase filters are formed by

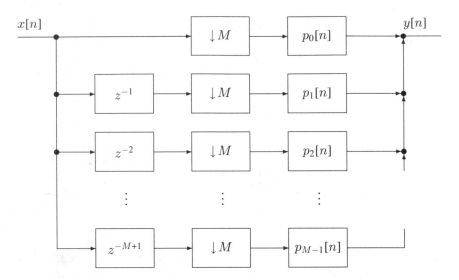

Figure 4.3.   A polyphase implementation of a decimator.

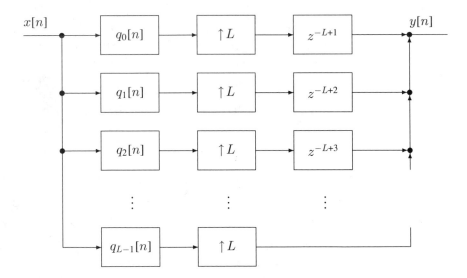

Figure 4.4.   A polyphase implementation of an interpolator.

downsampling the interpolation filter. If $g[n]$ is the lowpass interpolation filter, then the corresponding polyphase filters are

$$q_k[n] = g[nM + k]. \tag{4.15}$$

Polyphase filters can be either FIR or IIR filters. They often have the property that their magnitude responses approximate allpass filters, or they can also be strictly allpass. In the latter case, further computational savings can be achieved. Polyphase filters are used in many applications involving multirate filtering. Rate conversion is the simplest of these.

Efficient digital rate conversion is particularly important for oversampling A/D converters. These A/D converters sample the input at very high sampling rates to reduce the effects of aliasing. Later, the signal is decimated to the desired operating rate. The oversampling approach has the advantage that the major part of the anti-aliasing filtering is performed digitally, which avoids the need for expensive analog filters. It also permits linear-phase anti-aliasing filters, which would not be possible otherwise.

Polyphase filters are also used in analysis/synthesis systems for subband coding, which are discussed later in the chapter. For this application, one set of polyphase filters is typically used to realize more than one frequency selective filter simultaneously. Other applications where polyphase filters are employed are multirate implementations of general frequency selective filters. This topic is also discussed later in the chapter. These implementations can be particularly efficient.

EXERCISE 4.2.3.   **FIR Polyphase Filters**

Polyphase structures are efficient implementations for a broad class of multirate filters. This exercise examines the polyphase implementations associated with FIR filters and examines their properties.

(a) Use **f fdesign** to design a 128-point FIR lowpass filter with a cutoff frequency of $\pi/4$ using a Hanning window. Use **g dtft** to display the magnitude response of the filter. Sketch the plot.

(b) Now use **f lshift** and **f dnsample** to produce the four polyphase filters $p_0[n]$, $p_1[n]$, $p_2[n]$, and $p_3[n]$. For each, display and sketch the magnitude response and group delay using **g dtft**.

(c) If the polyphase filters were obtained from an ideal lowpass filter instead of the 128-point approximation, what would the group delays of the corresponding polyphase filters be? Determine the frequency responses for these ideal polyphase filters.   □

EXERCISE 4.2.4.   **More on FIR Polyphase Filters**

FIR polyphase filters often have approximately flat magnitude responses. How well this allpass characteristic is approximated depends on the filter order.

Design two FIR lowpass filters using the *Hamming* window option in **f fdesign**. Both should have a cutoff frequency of 1.57, which is approximately $\pi/2$. Let the first filter, $h_1[n]$, have length 16 and the second, $h_2[n]$, have length 15. Use **f lshift** and **f dnsample** to obtain two pairs of polyphase filters $p_0[n]$ and $p_1[n]$ and $p'_0[n]$ and $p'_1[n]$—the first pair derived from $h_1[n]$ and the second pair from $h_2[n]$. Note that these filters are obtained by downsampling by a factor of 2. Display and sketch the magnitude responses and pole/zero plots for each polyphase filter using **g polezero**. Explain why most of the polyphase filters have an approximate allpass appearance. Using the properties of linear phase filters, explain why one of the magnitude responses is zero at $\omega = \pi$. $\square$

EXERCISE 4.2.5. IIR Polyphase Filters

Equation (4.14) can be used to determine the polyphase filters from the impulse response of a prototype lowpass filter $H(z)$. However, if $H(z)$ is an IIR filter, (4.14) may not lead to polyphase filters that are rational functions.

To illustrate this point, consider the IIR M-to-1 decimation lowpass filter

$$H(z) = \frac{B(z)}{A(z)} = \frac{b[0] + b[1]z^{-1} + \ldots + b[N]z^{-N}}{1 + a[1]z^{-1} + \ldots + a[N]z^{-N}}$$

for which we want the polyphase filters $P_0(z)$, $P_1(z)$, ..., $P_{M-1}(z)$. To compute the polyphase filters using (4.14) would first require calculating the impulse response, $h[n]$, and then computing $P_0(z)$, $P_1(z)$, ..., $P_{M-1}(z)$ by downsampling. Since $h[n]$ is infinitely long, the polyphase filters will also be infinitely long, which is not attractive for implementation. If, however, $H(z)$ is of the form

$$H(z) = \frac{B(z)}{A(z^M)} = \frac{b[0] + b[1]z^{-1} + \ldots + b[N]z^{-N}}{1 + a[1]z^{-M} + \ldots + a[N]z^{-KM}}, \tag{4.16}$$

then the polyphase filters can be obtained by inspection. Specifically,

$$P_k(z) = \frac{\sum\limits_{m=0}^{M} b(Mm + k)z^{-m}}{1 + \sum\limits_{\ell=1}^{K} a[\ell]z^{-\ell}}. \tag{4.17}$$

Notice that all of the resulting polyphase filters have the same denominator. An arbitrary IIR lowpass filter $H(z) = B'(z)/A'(z)$ may be converted into the form of (4.16) by writing the system function in the following form [5], [11]:

$$H(z) = \frac{B'(z)}{A'(z)} \cdot \frac{\prod\limits_{k=1}^{M-1} A'(ze^{j2\pi k/M})}{\prod\limits_{k=1}^{M-1} A'(ze^{j2\pi k/M})} = \frac{B(z)}{A(z^M)}.$$

(a) Let $H(z)$ be the lowpass filter with cutoff frequency $\pi/2$ given by

$$H(z) = \frac{0.13 + 0.266z^{-1} + 0.36z^{-2} + 0.266z^{-3} + 0.13z^{-4}}{1 - 0.65z^{-1} + 1.09z^{-2} - 0.42z^{-3} + 0.18z^{-4}}.$$

Determine the polyphase filters $P_0(z)$ and $P_1(z)$ following the procedure suggested above. First use **f siggen** to create a file for $H(z)$ and **f convert** to obtain $B'(z)$ and $A'(z)$. Notice that, here, $M = 2$ and therefore $A'(e^{j2\pi k/M}z) = A'(-z)$, which can be realized by using **f cexp** with $\omega_0 = \pi$. Enter the letters *pi* instead of 3.14159 for the value of ω_0 in the function **f cexp**.

Use the **f convolve** and **f revert** functions to transform the numerator and denominator of $H(z)$ into the modified form $B(z)/A(z^M)$ shown in (4.16). The **f convert**, **f lshift**, **f dnsample**, and **f revert** can then be used to obtain the polyphase filters. Using **g polezero**, display and sketch the magnitude and pole/zero plots for $H(z)$, $P_0(z)$, and $P_1(z)$. Determine the stopband and passband ripple and transition width of $H(z)$ using **g dfilspec**.

(b) Consider the application of this IIR filter to a 2-to-1 decimator. How many multiplications and adds are required to implement the IIR decimator if the IIR filter is used in direct form? How many multiplications and adds are required for the IIR polyphase form? If an FIR filter were designed with comparable magnitude response characteristics, how many multiplications and adds would be required? The function **f eformulas** can be used to estimate the length of the FIR filter predicted by the Parks–McClellan formula.

You observe that the IIR filter is more efficient than the FIR filter in this case. However, the polyphase form does not seem to yield improvement over direct form implementation. Because the conversion of $H(z)$ into the form of (4.16) increases the filter order, this approach to IIR polyphase implementation does not, unfortunately, result in reducing the computational load. A more efficient approach is discussed in the next exercise.   □

EXERCISE 4.2.6. **Allpass Polyphase Filters**

When the polyphase filters corresponding to lowpass and highpass filters are allpass, highly efficient implementations are often possible. Let $P_k(z)$ be the kth polyphase filter in a multirate system. In this exercise, $P_k(z)$ will be constrained to have the form

$$\frac{1}{M} \prod_{m=1}^{M} \frac{z^{-1} - \alpha_m}{1 - \alpha_m z^{-1}}, \tag{4.18}$$

where the α_m's are real coefficients.

(a) Consider the multirate filter, $H(z)$, which has the polyphase expansion

$$H(z) = P_0(z^2) + z^{-1} P_1(z^2) = \frac{z^{-2} - \alpha_0}{1 - \alpha_0 z^{-2}} + z^{-1} \frac{z^{-2} - \alpha_1}{1 - \alpha_1 z^{-2}}. \tag{4.19}$$

The coefficients α_0 and α_1 for $H(z)$ can be determined iteratively so that $H(e^{j\omega})$ approximates a desired frequency response using optimization programs that minimize a frequency domain error expression. Such programs were used to find the allpass coefficients given below so that $H(z)$ would approximate a lowpass filter with a cutoff frequency of $\pi/2$.

(i) Using the *create file* option in **f siggen**, create a file to represent the IIR filter, $P_0(z^2)$. Notice that the numerator and denominator filter orders will be 2, the sequence values will be real, and the starting point will be zero. For the coefficients use

$$
\begin{aligned}
b_0 &= 0.1413 \\
b_1 &= 0 \qquad\qquad a_1 = 0 \\
b_2 &= 1 \qquad\qquad a_2 = 0.1413
\end{aligned}
$$

Also create a similar file for $z^{-1}P_1(z^2)$. In this case, let the numerator order be 3, the denominator order be 2, and the starting point be zero. The coefficients of the filter are

$$
\begin{aligned}
b_0 &= 0 \\
b_1 &= 0.5898 \qquad a_1 = 0 \\
b_2 &= 0 \qquad\qquad a_2 = 0.5898 \\
b_3 &= 1
\end{aligned}
$$

This will also account for the z^{-1} delay term.

Use the **f par** function to combine the two polyphase filters into the single filter, $H(z)$. Display the magnitude response using **g dfilspec** and record the transition width and stopband ripple of the filter. Assume that the passband deviation is 0.01. Sketch the magnitude response.

(ii) Assuming that the filter is realized using the multirate polyphase structure shown in Figure 4.3 and the allpass filters are implemented efficiently using difference equations of the form shown in (4.2), determine the number of multiplications and additions required for each output sample.

(iii) Use **f eformulas** to determine the length of a comparable Parks–McClellan FIR filter. How many multiplications and additions are required to implement the Parks–McClellan filter? Notice the savings achieved by using allpass polyphase filters.

(iv) Now consider

$$
H(z) = \frac{z^{-2} + 0.4}{1 + 0.4z^{-2}} + z^{-1}.
$$

Repeat part (i) based on this new filter. Note that only half the number of multiplications are required. Also note that the magnitude response characteristics are not quite as good due to the lower order of the filter. Evaluate the number of multiplications and adds required to implement a Parks–McClellan filter with comparable magnitude response specifications using **f eformulas**.

(b) Consider the multirate lowpass filter, $H(z)$, with cutoff frequency $\pi/4$. The filter is expressible in terms of polyphase filters and has the form

$$
H(z) = P_0(z^{-4}) + z^{-1}P_1(z^{-4}) + z^{-2}P_2(z^{-4}) + z^{-3}P_3(z^{-4}).
$$

Each polyphase filter is of the form shown in (4.18) with $M = 1$. As in the previous part, use the *create file* option in **f siggen** to create files for each of the four allpass polyphase filters, $P_0(z)$, $P_1(z)$, $P_2(z)$, and $P_3(z)$, where the corresponding allpass coefficients for the first three filters are -0.2076, -0.4257, and -0.6772. $P_4(z)$ is a delay of three samples. In this case, the numerator and denominator filter orders should be seven for each polyphase filter. These filters incorporate the delay factors.

For $P_0(z)$, enter:

$b_0 = 0.2076$

$b_1 = 0$ $a_1 = 0$

$b_2 = 0$ $a_2 = 0$

$b_3 = 0$ $a_3 = 0$

$b_4 = 1$ $a_4 = 0.2076$

$b_5 = 0$ $a_5 = 0$

$b_6 = 0$ $a_6 = 0$

$b_7 = 0$ $a_7 = 0$

For $P_1(z)$, enter:

$b_0 = 0$

$b_1 = 0.4257$ $a_1 = 0$

$b_2 = 0$ $a_2 = 0$

$b_3 = 0$ $a_3 = 0$

$b_4 = 0$ $a_4 = 0.4257$

$b_5 = 1$ $a_5 = 0$

$b_6 = 0$ $a_6 = 0$

$b_7 = 0$ $a_7 = 0$

For $P_2(z)$, enter:

$b_0 = 0$

$b_1 = 0$ $a_1 = 0$

$b_2 = 0.6772$ $a_2 = 0$

$b_3 = 0$ $a_3 = 0$

$b_4 = 0$ $a_4 = 0.6772$

$b_5 = 0$ $a_5 = 0$

$b_6 = 1$ $a_6 = 0$

$b_7 = 0$ $a_7 = 0$

For $P_3(z)$, enter:

$b_0 = 0$

$b_1 = 0$ $a_1 = 0$

$b_2 = 0$ $a_2 = 0$

$b_3 = 1$ $a_3 = 0$

$b_4 = 0$ $a_4 = 0$

$b_5 = 0$ $a_5 = 0$

$b_6 = 0$ $a_6 = 0$

$b_7 = 0$ $a_7 = 0$

Use the **f par** function successively to combine the four polyphase filters into a single filter, $H(z)$. Display and sketch the magnitude response of $H(z)$ using **g dtft**. Observe that there are spikes in the stopband region of the magnitude response. These artifacts are typical of narrowband lowpass filters based on allpass decompositions. If the filter order is increased, however, the severity of these spikes diminishes.

Assuming that the filter is realized using the multirate polyphase structure shown in Figure 4.3 and the allpass filters are implemented efficiently using difference equations of the form shown in (4.2), determine the number of multiplications and adds required per output sample.   □

EXERCISE 4.2.7. **Allpass Polyphase Decomposition**

The previous exercise showed that filter implementations based on allpass polyphase filters could be highly efficient. These filters can be designed by iterative techniques that minimize weighted frequency domain error expressions. For filters with a cutoff frequency $\pi/2$, a convenient noniterative method is available. It is based on decomposing classical IIR filters into two allpass polyphase filters, each having the form of (4.16) with $M = 2$.

When odd order digital lowpass Butterworth and elliptic filters are designed with passband and stopband cutoff frequencies that are symmetric about $\pi/2$ and $\delta_p = 0.5(1 - \sqrt{1 - \delta_s^2})$, the resulting filters will have the form of (4.16) with $M = 2$.

(a) Consider the design of a digital elliptic filter with $\omega_p = 0.45\pi$, $\omega_s = 0.55\pi$, $\delta_p = 0.0016$, and $\delta_s = 0.08$. Notice that these specifications satisfy the conditions stated above. This elliptic filter may be designed as follows:

(i) Use **f aelliptic** with $\Omega_s = 1.18887$, $\Delta_p = 0.0031949$, $\Delta_s = 0.07987$, and $N = 5$ to obtain the lowpass prototype.

(ii) Use **f bilinear** with $\beta = 2$ to convert the analog prototype to digital form.

(iii) Use **f dtransform** with $c_0 = 0.294358$, $c_1 = 1$, $c_2 = 0$, $d_0 = 1$, $d_1 = 0.294358$, $d_2 = 0$ to center the cutoff frequency so that the response will be symmetric about $\pi/2$.

(iv) Use **f gain** with a gain factor of 1.0016 to scale the filter amplitude.

Design the filter and examine it using **g polezero**. Sketch the pole/zero plot and magnitude response. Observe that the poles occur as complex conjugate roots on the $j\omega$-axis, which implies that the denominator contains nonzero coefficients only for even powers of z^{-1}. Use the *list pole/zero* option to display the numerator and denominator polynomials. Observe that the denominator polynomial is a polynomial in z^{-2} and thus $H(z)$ may be decomposed into two polyphase filters. Determine and record these polyphase filters. How many multiplications and adds are required per polyphase filter output?

(b) Construct the polyphase filters, $P_0(z)$ and $P_1(z)$, using **f convert, f dnsample, f lshift**, and **f revert**. Display and sketch the pole/zero plot for each using **g polezero**. Adjust the circle size for a good display and examine the pole/zero plot carefully. Observe that there is pole/zero cancellation, which implies that the order of the polyphase filters may be reduced. Use the *list pole/zero* option to determine the allpass polyphase filter coefficients α_0 and α_1, where

$$P_0(z) = \frac{z^{-1} - \alpha_0}{1 - \alpha_0 z^{-1}} \quad P_1(z) = \frac{z^{-1} - \alpha_1}{1 - \alpha_1 z^{-1}}$$

and record these values. Recognizing that these are allpass polyphase filters, how many multiplications and adds are required to calculate each polyphase filter output sample?

The procedure outlined in this exercise provides a fairly simple way to design allpass polyphase filters for 2-to-1 decimation and interpolation. The elliptic filter design program given in the software, however, is not the best to use for this purpose because it does not allow you to set the stopband ripple precisely. The function **f aelliptic** allows you to set the passband ripple, stopband cutoff frequency, and filter order. It designs the elliptic filter with smallest stopband ripple. Other elliptic filter design routines allow for both ripples, the center cutoff frequency, and the order to be specified and design the elliptic filter with the smallest transition region. Design programs of this type are more convenient for designing these filters.   □

4.2.2 Subband Decompositions

Subband coding is a popular technique for compressing speech, images, and video signals that has motivated much of the recent research in multirate systems. With this technique, the input signal is first split into several bandpass components using several filters called a *filter bank* or, more precisely, an *analysis filter bank*. The analysis filter bank typically contains a lowpass filter, bandpass filters, and a highpass filter that collectively cover the frequency spectrum without any spectral gaps. The outputs of the filters are downsampled to their respective Nyquist rates and the downsampled signals, known as *subbands*, are then coded for transmission or storage. The number of bits used to code each subband is varied based on the perceptual importance of that subband. It has been shown for audio and visual signals that coding subbands generally results in better performance than coding the input signal directly. The reconstruction procedure, which is called *synthesis*, consists of upsampling the subbands, filtering them with an appropriate filter, and then summing the results together. This part of the subband coding system is called the *synthesis filter bank* or *synthesis system*.

There are several issues that arise when designing subband coding systems. First, the analysis and synthesis filters must be designed to have good magnitude response characteristics to isolate the spectral components of the signal. In some cases, having a linear phase filter response characteristic is thought to be important. Second, the number of bands and the widths of the individual bands must be chosen carefully. These choices are influenced by computational limitations, system delay constraints, integer band sampling constraints, input signal characteristics, and other factors. Third, in the absence of coding, the analysis and synthesis filter banks should not introduce any significant distortion. Because decimation introduces aliasing in the output, the filter banks must be designed carefully in order to avoid these and other distortions.

EXERCISE 4.2.8. **A Two-Band Filter Bank**

In the earlier discussion, it was shown that a decimator could be implemented using polyphase filters. With a relatively small increase in computational complexity, 2-to-1 decimators can be modified to create two-band filter banks.

Consider the lowpass filter, $H_0(z)$, of the form

$$H_0(z) = P_0(z^2) + z^{-1}P_1(z^2).$$

A highpass filter $H_1(z)$ can be constructed by subtracting rather than adding the polyphase filters.

(a) Use **f fdesign** to design a length 32 FIR lowpass filter with a Hamming window. Let the cutoff frequency of the filter be $\pi/2 = 1.5708$. Use **f dnsample** and **f lshift** to obtain the two polyphase filters $P_0(z)$ and $P_1(z)$. Note that in creating $P_1(z)$, you should first shift $H_0(z)$ by -1 and then downsample. Next use **f upsample** to form $P_0(z^2)$ and $P_1(z^2)$. Shift $P_1(z^2)$ by 1 using **f lshift** to form $z^{-1}P_1(z^2)$. Using **f add** and **f subtract** compute

$$H_0(z) = P_0(z^2) + z^{-1}P_1(z^2)$$

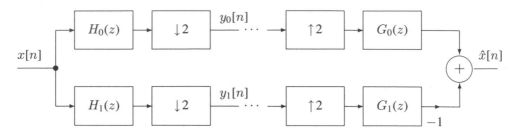

Figure 4.5. Two-band analysis and synthesis filter bank.

and

$$H_1(z) = P_0(z^2) - z^{-1}P_1(z^2).$$

Display their magnitude responses using **g dtft** and sketch the plots.

(b) Show analytically why $H_1(z)$ is a highpass filter.

(c) Draw the block diagram for the efficient polyphase realization of a two-band analysis filter bank.

(d) A two-band synthesis filter bank has two input signals; $y_0[n]$, corresponding to the lowpass subband and $y_1[n]$, corresponding to the highpass subband. These inputs, $y_0[n]$ and $y_1[n]$, are upsampled by a factor of 2, and filtered with lowpass and highpass filters, respectively, prior to being summed as shown in Figure 4.5. Draw a block diagram for the polyphase realization of the corresponding two-band synthesis filter bank.   □

EXERCISE 4.2.9.   QMF Filter Bank

Consider the two-band analysis and synthesis filter banks in cascade shown in Figure 4.5. Let $x[n]$ be the input and $\hat{x}[n]$ be the output. If $\hat{x}[n] = x[n - \Delta]$, the two-band system is said to be *exactly* or *perfectly* reconstructing. In this case the input and output are identical except for a delay of Δ samples. Write a macro that implements the two-band analysis/synthesis system shown in Figure 4.5. This can be done using the functions **f convolve**, **f dnsample**, **f upsample**, **f gain**, and **f subtract**.

(a) Use **f fdesign** to design an 8-point lowpass filter, $H_0(z)$, and an 8-point highpass filter, $H_1(z)$, both with a cutoff frequency of $\pi/2$ using a rectangular window. Generate the input signal $x[n] = \delta[n] + \delta[n - 1]$ using **f siggen**. Let $G_0(z) = H_0(z)$ and $G_1(z) = H_1(z)$. Use $x[n]$ as the input to your macro and compute its output. If the system is exactly reconstructing, the input will appear at the output delayed by seven samples. Display and sketch $\hat{x}[n]$ using **g view**. Notice that $\hat{x}[n]$ resembles the input but some distortion has been introduced. In addition, observe that the output has been scaled by a factor of $1/2$. Explain why.

(b) Now consider a different lowpass filter, $H_0(z) = H_a(z)$, whose coefficients are given in Table 4.1. Use the *create file* option in **f siggen** to put these coefficients in a file. You may wish to save the file for $H_a(z)$ as it will be used again in a later exercise.

Table 4.1. $H_a(z)$ coefficients for an 8-length QMF (design by Johnston [6]). $H_b(z)$ coefficients for an optimal equiripple perfect reconstruction filter (Smith and Barnwell [10]). $H_c(z)$ coefficients for an 8-length wavelet generating filter.

$H_a(z)$	$H_b(z)$	$H_c(z)$
0.009387150	0.034897551	−0.162898020
−0.070651830	−0.010983018	−0.505463067
0.069428270	−0.062864534	−0.446096523
0.489980800	0.223907702	0.019777670
0.489980800	0.556856947	0.132245320
0.069428270	0.357976275	−0.021807744
−0.070651830	−0.023900269	−0.023250397
0.009387150	−0.075940958	−0.007493520

Let $H_1(z) = H_0(-z)$, $G_0(z) = H_0(z)$, and $G_1(z) = H_1(z)$. Recognize that the time domain relationship between $H_0(z)$ and $H_1(z)$ is

$$h_1[n] = (-1)^n h_0[n] = e^{j\pi n} h_0[n].$$

As a result the **f cexp** function with $\omega_0 = \pi$ may be used to obtain $H_1(z)$ from $H_0(z)$. Enter the letters *pi* for the value of ω_0 in the function **f cexp** to obtain $H_1(z)$. Use **g dtft** to display the magnitude responses for $H_0(z)$ and $H_1(z)$. They should be lowpass and highpass filters, respectively. These filters are called *quadrature mirror filters* or *QMFs* [4]. This particular filter was designed by Johnston using an iterative optimization procedure [6]. A useful set of Johnston's QMFs with different lengths and characteristics is available in [6].

Use $x[n]$ as the input to your macro and compute the output. Display and sketch the output using **g view**. Notice that the distortion is very small.

(c) QMFs have the property that aliasing is canceled in the synthesis filter bank. Write an equation for $\hat{X}(e^{j\omega})$ in terms of $X(e^{j\omega})$ and the filters $H_0(e^{j\omega})$ and $H_1(e^{j\omega})$ and show that aliasing is, in fact, canceled in the synthesis filter bank. QMFs are designed to have good reconstruction properties and are used in many subband coding systems for compressing speech, images, and video signals.   □

EXERCISE 4.2.10.   Tree-Structured Two-Band Decompositions

Two-band systems can be cascaded in tree structures to split a single signal into many subband components successively. For example, an eight-band decomposition (analysis system) can be produced by first splitting the input signal into two bands, then splitting each of those into two additional bands for a total of four bands and finally splitting the four again for a total of eight. The original signal can be reconstructed by merging the subbands two at a time using the two-band synthesis filter bank shown in Figure 4.5.

An *octave band* tree structure is one type of tree in which only the lowpass band is split successively. Such an octave band analysis system is shown in Figure 4.6.

(a) Write a macro that implements a four-band octave band analysis system using the quadrature mirror filter (QMF), $H_a(z)$, given in Table 4.1, where $H_0(z) = \sqrt{2} H_a(z)$

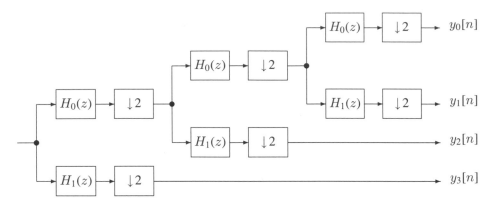

Figure 4.6.   A Four-Band Octave Band Tree Structure.

and $H_1(z) = \sqrt{2}H_a(-z)$. Notice that $h_1[n]$ can be obtained using **f cexp** with $\omega_0 = \pi$ because $h_1[n] = (-1)^n h_a[n] = e^{(j\pi n)} h_a[n]$. Enter the letters *pi* for the value of ω_0 in the function **f cexp** to obtain $h_1[n]$. The function **f gain** can be used to scale the filter by a factor of $\sqrt{2}$. Use the full decimal equivalent $\sqrt{2} = 1.41421356237$ when entering the value of $\sqrt{2}$ in **f gain** as this will help you achieve the best results. Your macro should be composed of the functions **f convolve** and **f dnsample**.

Next write a macro that implements the synthesis or reconstruction system for the octave band decomposition. For the synthesis system (see Figure 4.5), $g_0[n] = h_0[n]$ and $g_1[n] = h_1[n]$. This macro will take each of the four bands, upsample and filter them successively, and accumulate the results to form the output. If the analysis and synthesis systems are working properly, the output will be identical to the input but shifted in time by an amount equal to the system delay.

The subband signals $y_2[n]$ and $y_3[n]$ will have to be delayed appropriately in the synthesis system to compensate for the delay introduced in the lower frequency bands. This compensation delay in the $y_2[n]$ and $y_3[n]$ bands is necessary so that the subband channels will all be time aligned before summing. The synthesis macro should contain the functions **f lshift**, **f upsample**, **f convolve**, and **f subtract**. What is the numerical value of the compensation delay for the $y_2[n]$ band? What about the delay for the $y_3[n]$ band? Save these macros and filters. They will also be used in the following exercise.

(b) Let the input to the analysis system be $x[n] = \delta[n] + \delta[n-1]$. Evaluate the subbands and reconstruct the signal using your synthesis macro. Display the output using **g view**. What is the overall delay experienced by the input (i.e., the system delay)? Shift the output signal back to zero using **f lshift** so that it is time aligned with the input. Using **f snr**, compute and record the signal-to-noise ratio. If your system is working properly, the SNR should be above 40dB.

Octave band structures of this type are used extensively in subband coding of speech, images, and video. Discrete wavelet decompositions, which have been introduced more recently, utilize this same octave band structure. The generating filters employed in the wavelet decompositions are similar to the QMFs but have some differences as we shall see in the following exercise.   □

EXERCISE 4.2.11. Perfect Reconstruction Filter Banks

In this exercise, we continue the investigation of octave band tree structures. Therefore, you should work the previous exercise prior to doing this one.

The performance of an octave band tree is largely dependent on the filters used in the analysis and synthesis systems. Three representative filters are given in Table 4.1. The first is a Quadrature Mirror Filter (QMF) design by Johnston [6]. QMFs were the first known filters capable of reconstruction without aliasing distortion. From the previous exercise, record the SNR obtained using the QMF with input $x[n] = \delta[n] + \delta[n-1]$ using your analysis and synthesis macros.

(a) Another two-band filter class is called *perfect reconstruction* (PR). When applied in a two-band system or tree structure, they allow the input to be reconstructed without any filtering distortion and are accurate to within the numerical precision limits of the digital processor. Within this general class of PR filters are two interesting types that we will examine briefly. The first one, which will be examined in this part of the exercise, is an optimal equiripple PR filter, $H_b(z)$, which was designed using the method in [9]. Its coefficients are given in Table 4.1. Using the *create file* option in **f siggen** create a file for this filter. The corresponding analysis and synthesis filters are

$$h_0[n] = \sqrt{2}h_b[n]$$
$$h_1[n] = (-1)^n h_0[L - n - 1]$$
$$g_0[n] = h_0[L - n - 1]$$
$$g_1[n] = (-1)^n h_0[n], \tag{4.20}$$

where L is the filter length, which in this case is 8. Notice that these filters can be constructed using **f reverse**, **f lshift**, and **f cexp** (by specifying the letters *pi* for the value of ω_0). The scale factor $\sqrt{2} = 1.41421356237$ can be added by using the **f gain** function.

Use these filters in your macro and evaluate the output for the input $x[n]$. As before, compute and record the SNR using **f snr**. Save the filter file for $h_b[n]$, as it will be used in the next exercise.

(b) The second type of PR filter we will examine is a wavelet generating filter, $h_c[n]$. This filter was designed following the same approach used to design $h_b[n]$. However, in this case the design was based on a maximally flat FIR filter instead of an equiripple filter. The coefficients for this filter are given in Table 4.1. The filter assignments given in (4.20) should be used to form the individual analysis and synthesis filters. As before, create a file for this filter, construct the analysis and synthesis filters, and use them in your macros. Shift the output back in time to account for the system delay and evaluate and record the SNR using **f snr** as before. Save the $h_c[n]$ file for the next exercise.

(c) Based on reconstruction quality, how do the two PR filters $h_b[n]$ and $h_c[n]$ compare to the QMF $h_a[n]$? Are the differences you observe in the plots dramatic?   □

EXERCISE 4.2.12.   **Regularity and Discrete Wavelets**

This exercise continues the discussion on octave band trees that began in the previous two exercises. Read through both of the previous exercises before beginning this one.

Table 4.1 lists coefficients for three types of filters: a QMF, an optimal equiripple PR filter, and a discrete wavelet generating function [12]. Files for each of these filters should be created if they have not been created already.

(a) Each of the subbands of an octave band tree has a direct form filter associated with it. That filter can be obtained using the multirate identities discussed in Exercise 4.2.2. In particular, the direct form filter corresponding to the $y_0[n]$ band is $F_0(z) = H_0(z)H_0(z^2)H_0(z^4)$. For the $y_1[n]$, $y_2[n]$, and $y_3[n]$ bands, the direct form filters are $F_1(z) = H_0(z)H_0(z^2)H_1(z^4)$, $F_2(z) = H_0(z)H_1(z^2)$, and $F_3(z) = H_1(z)$, respectively. If a discrete wavelet generating function such as $H_c(z)$ is used in the octave tree, the direct form filters $F_1(z)$, $F_2(z)$, and $F_3(z)$ are called *discrete wavelets*. Using the functions **f upsample** and **f convolve** and $H_c(z)$, compute the discrete wavelets. Display and sketch them using **g view**. Also display and sketch their magnitude response using **g dtft**. Observe that wavelets with short time durations have large passband regions and ones with long time durations have short bandwidths. This is a feature that distinguishes octave trees (and discrete wavelet transforms) from classical uniform band decompositions such as the short-time Fourier transform.

(b) A discrete wavelet has the property that it is *regular*. By this we mean that the wavelet, $F_1(z)$, and the scaling function $F_0(z)$ converge to a smooth time function as the number of band splits goes to infinity.

As a practical matter for determining regularity, it is generally sufficient to evaluate six or seven octave splits in order to see whether $F_0(z)$ is converging to a smooth function.

To explore this concept, write a macro, **reg.bat**, to compute

$$H_0(z)H_0(z^2)H_0(z^4)H_0(z^8)H_0(z^{16})H_0(z^{32}),$$

using **f upsample** and **f convolve**. Evaluate $F_0(z)$ for the generating function $H_c(z)$. Display and sketch the filter in the time domain using **g look**. Now repeat the same for the QMF, $H_a(z)$, and the optimal equiripple PR filter, $H_b(z)$. Which has the highest degree of regularity (or smoothness in the time domain)?

(c) Carefully examine the DTFT magnitude and log magnitude for each of the macro outputs in part (b) using **g dtft**. Which has the best magnitude response characteristics in terms of transition width and peak attenuation? Which has the worst? Examine the pole/zero plots for $h_a[n]$, $h_b[n]$, and $h_c[n]$. It has been shown that having zeros at $z = -1$ in the analysis filter (or wavelet generating function) is important for regularity. Is this consistent with your observation?   □

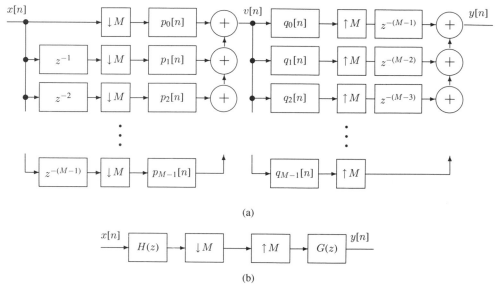

(a)

(b)

Figure 4.7. Multirate structures for lowpass and bandpass filters. a) polyphase implementation. b) direct multirate implementation.

4.2.3 Multirate Frequency Selective Filters

Frequency selective filters, such as lowpass, highpass, bandpass, and bandstop filters, can be implemented as multirate filters. In cases where the filters have either very broad or very narrow passbands, impressive reductions in the number of required computations are possible. The basic multirate structure, which is shown in Figure 4.7, involves a decimator in cascade with an interpolator, both of which are usually implemented in polyphase form. To investigate the computational advantages of this approach, assume for the moment that we wish to design an FIR lowpass filter with a nominal cutoff frequency of π/M and maximum passband and stopband ripples of δ_p and δ_s, respectively. This can be done using the multirate realization shown in Figure 4.7 in which $H(z)$ and $G(z)$ are FIR lowpass filters of length L. The filters $H(z)$ and $G(z)$ must be designed so that their ripples, $\hat{\delta}_p$ and $\hat{\delta}_s$, which are the same for both filters, satisfy the relationship:

$$\hat{\delta}_p \leq \delta_p/2 \qquad (4.21)$$
$$\hat{\delta}_s \leq \delta_s. \qquad (4.22)$$

The stopband cutoff frequencies should be π/M.

To estimate the amount of computation involved, we can count the number of multiplies and adds. The approximate length of each polyphase filter is L/M. Therefore, for each polyphase filter approximately L/M multiplications and additions are required to generate one polyphase output sample at the lower sampling rate. In both the decimation and interpolation sections, there are a total of $2M$ polyphase filters, which means that approximately $2L$ multiplications and additions must be performed at the lower rate.

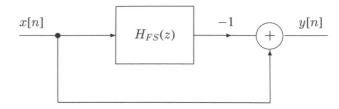

Figure 4.8. Multirate structure for implementing highpass and bandstop filters.

Because there is only one sample of $v[n]$ (see Figure 4.7) for every M samples of the output $y[n]$, the number of multiplications and additions per output is approximately $2L/M$. Notice that a direct implementation of an FIR lowpass filter of length L requires approximately L multiplications and additions for each output sample.

We can use the computational gain ratio, G_c, to measure efficiency. This ratio is defined as the number of multiplications required by the direct implementation divided by the number of multiplications required by the multirate implementation. This yields

$$G_c = \frac{L}{2L/M} = \frac{M}{2}.$$

The amount of computation is directly related to M, and savings over the direct form implementations are achieved for $M > 2$.

This multirate structure can be used to realize a variety of constant band filters efficiently. Bandpass filters, for example, can be realized using the structure in Figure 4.7. Highpass and bandstop filters can be constructed by using the structure in Figure 4.8, where $H_{FS}(z)$ is a multirate lowpass or bandpass filter. The greatest computational gains are generally achievable when the passbands are either very narrow or very broad.

EXERCISE 4.2.13. **Multirate FIR Lowpass Filters**

Multirate lowpass filters were discussed in the section introduction. They have the form shown in Figure 4.7. This exercise is concerned with implementing and evaluating such a filter. Begin by designing a lowpass prototype filter of length 55 and cutoff frequency 0.65 using **f ideallp**. Window the ideal filter with a Kaiser window with $\alpha = 3$ using **f kaiser** and **f multiply**. This filter will be used for $H(z)$ and $G(z)$ in the multirate filter. Use **f gain** to scale $G(z)$ by 4 to account for the gain factor of $1/4$ in the 4-to-1 decimator.

(a) In practice, we would usually implement the multirate filter in polyphase form as shown in Figure 4.7, in which case only about $2L/M \approx 28$ multiplications per output sample would be required. However, for our purpose here, it will be sufficient to implement the decimator and interpolator directly using **f convolve**, **f dnsample**, and **f upsample**. Write a macro **mrate.bat** that implements this multirate lowpass filter. Use the impulse signal provided in the file **impulse** as the input to the macro and compute the impulse response $h[n]$ of the multirate filter. Display its magnitude response using **g dfilspec** and determine the passband and stopband ripples and lower and upper cutoff frequencies. Sketch the plot. Save $h[n]$ and **mrate.bat**, as they will be used again in the next exercise.

(b) To assess the computational advantages of this multirate filter now that the cutoff frequencies and ripple specifications are known, consider finding the lengths of direct form filters that satisfy these same constraints. Use the **f eformulas** function to estimate the lengths of the best Kaiser window and Parks–McClellan filters that meet the tolerances found in part (a). Determine the computational gain ratios G_c based on a Kaiser filter and a Parks–McClellan filter. Note the savings in arithmetic.

(c) Since the decimation factor determines the computational efficiency of the multirate implementation, it is easy to see that narrowband filters tend to yield very efficient multirate realizations. However, we can also obtain high efficiency for very wide band filters. Note that if $h[n]$ is an odd-length zero-phase narrowband lowpass filter with a cutoff frequency of ω_0, then $(-1)^n h_0[n]$ is a narrowband highpass filter with cutoff frequency $\pi - \omega_0$ and $1 - (-1)^n h_0[n]$ is a wideband lowpass filter with cutoff frequency $\pi - \omega_0$. Describe how you would implement an efficient wide band lowpass filter with a cutoff frequency of $7\pi/8$. $\qquad\square$

EXERCISE 4.2.14. **Multirate Filters Are Time Varying**

In the discussion thus far, multirate filters have been treated as substitutes for linear time invariant or LTI filters. Although the implication is that they are equivalent, in actuality they are not. Multirate filters of this form are linear but not time invariant. More precisely, they are linear *periodically time varying* filters with a period of M. The primary reason for the loss of time invariance is the aliasing introduced in the decimation process. However, when aliasing is forced to be small by using decimation filters with near ideal magnitude responses, the equivalent filters are approximately time invariant.

To illustrate this point, we will use the macro **mrate.bat** that was defined in Exercise 4.2.13 part (a). If you have not already done so, write this macro now.

(a) Design a 17 length lowpass filter with cutoff frequency $\pi/4 = 0.7854$ using the *rectangular* window in **f fdesign**. Use this filter for $H(z)$ and $G(z)$ in the macro. Compute the impulse response of the multirate filter (as described in the previous exercise) and save it in the file **h0**. Next compute the shifted impulse, $\delta[n - 1]$ using **f lshift** and evaluate the shifted impulse response of the multirate filter. Store this output in the file **h1**. If the multirate filter were linear and time invariant, the impulse responses in **h0** and **h1** would be identical except for a time shift. However, since the multirate filter is time varying, there are differences. Display **h0** and **h1** using **g view2** and sketch the plots.

(b) Generate the shifted impulses, $\delta[n - 2]$, $\delta[n - 3]$, and $\delta[n - 4]$. Compute the shifted impulse responses for each and store them in the files **h2**, **h3**, and **h4**, respectively. Using **g view2**, compare each of these with **h0**. Which is identical to **h0** (to within a shift factor)?

The sequence of responses, obtained by inputting shifting impulses to the multirate filter, is periodic. When systems exhibit this kind of behavior, they are said to be *periodically time varying*. $\qquad\square$

EXERCISE 4.2.15.   **Multistage Multirate Lowpass Filter**

Multistage implementations of digital filters are composed of cascades of decimators and interpolators. This approach can result in additional savings in computation. This exercise will look at a narrowband four-stage multirate FIR lowpass filter.

(a) Begin by designing a lowpass prototype filter, $h[n]$, using the Parks–McClellan algorithm, which is implemented in the function **f pksmcc**. Specify $\delta_s = \delta_p = 0.001$, $\omega_p = 0.25\pi$, and $\omega_s = 0.5\pi$. The length of $h[n]$ should be 26. Use $h[n]$ in each stage of the decimator and $2h[n]$ in each stage of the interpolator.

Write a macro that will implement a four-stage multirate lowpass filter. For convenience, use a direct form implementation of the decimators and interpolators in the macro instead of the polyphase form. Note, the macro will contain a cascade of four 2-to-1 decimators followed by a cascade of four 2-to-1 interpolators. This can be done using the three functions **f convolve**, **f dnsample**, and **f upsample**. Use the filter $h[n]$ in each decimator and interpolator stage.

(b) Use an impulse, which is given in the file **impulse** as the input to the macro. This will produce the impulse response, $f[n]$, of the four-stage multirate filter. Display and sketch the magnitude response of $f[n]$ using **g dfilspec**. Then determine the passband ripple, stopband ripple, and transition band. Use **f eformulas** to determine the minimum length Parks–McClellan filter that satisfies the passband, stopband, and transition width specifications found for $f[n]$.

(c) Sketch the block diagram for a polyphase implementation of a four-stage multirate filter. Notice that the form will be the same as that shown in Figure 4.7 except that there will be four cascaded 2-to-1 decimators and four cascaded 1-to-2 interpolators instead of a single M fold decimator and interpolator as shown in the figure.

(d) Assuming an efficient polyphase implementation for the decimators and interpolators, determine the approximate number of multiplications and additions required to implement the four-stage multirate filter.

How many multiplications and additions would be required for a direct form implementation of an FIR filter that satisfies the same frequency specifications as $f[n]$, but which is designed using the Parks–McClellan algorithm?   □

In general, the efficiency associated with multistage implementations increases as the passband of the lowpass filter is made narrower.

EXERCISE 4.2.16.   **Multirate Bandpass Filters**

The implementation that we saw in Figure 4.7 for implementing a multirate lowpass filter can also be used to implement multirate bandpass filters. In this exercise, you will design a multirate bandpass filter and evaluate its performance.

The first step is to design a 100-point FIR lowpass prototype filter, $h_0[n]$, with cutoff frequency 0.2 using the *Hamming* window option in **f fdesign**. Convert this prototype into a bandpass filter $h_b[n]$, where

$$h_b[n] = e^{j\omega_0}h_0[n] + e^{-j\omega_0}h_0[n]$$

and $\omega_0 = 5\pi/8 = 1.178097$. This can be done using the **f cexp** and **f add** functions.

Display and sketch the magnitude response of $h_b[n]$ using **g dtft** and verify that it is a bandpass filter. This filter will be used for $H(z)$ and $G(z)$ in the multirate filter structure shown in Figure 4.7 with decimation factor, $M = 4$. Use **f gain** to scale $G(z)$ by 4 to account for the $1/4$ gain factor in the 4-to-1 decimation.

Write a macro that implements the multirate bandpass filter. If you have already done Exercise 4.2.13, you already have this macro available—**mrate.bat**. For simplicity, use the nonpolyphase form to implement the decimator and interpolator with **f convolve, f dnsample**, and **f upsample**. Generate an impulse using **f siggen** and use it as the input to the macro. Compute the impulse response, $h[n]$, of the multirate filter. Display its magnitude response using **g dfilspec** and determine the passband and stopband ripples. Sketch the plot.

This example illustrates that bandpass filters can be implemented using efficient multirate structures. The amount of efficiency is directly related to the magnitude of the decimation factor M.   □

EXERCISE 4.2.17.   **Bandpass Filters and the Integer Band Constraint**

When ideal bandpass filters have upper and lower cutoff frequencies that are integer multiples of π/M, where M is an integer, downsampling by M will relocate the passband to the spectral region between $-\pi$ and π without creating aliasing. In these cases, the bandpass filter is said to satisfy the *integer band sampling* condition. When the filter's cutoff frequencies do not satisfy this condition, downsampling usually produces undesirable spectral overlap (aliasing). This phenomenon, which creates a constraint for multirate filter implementations, is examined in this exercise. A solution is also considered that reduces the severity of the constraint.

(a) Design a 100-point FIR lowpass baseband filter $h_0[n]$, using the *Hamming window* option in **f fdesign** with cutoff frequency 0.2. Next design the bandpass filter $h_b[n]$, where

$$h_b[n] = e^{j\omega_0} h_0[n] + e^{-j\omega_0} h_0[n] \qquad (4.23)$$

and $\omega_0 = 5\pi/8 = 1.178097$. This can be done using the **f cexp** and **f add** functions.

Now design another bandpass filter, $h_a[n]$, using (4.23) but with $\omega_0 = 1.47$. Display the magnitude responses for the bandpass filters, $h_a[n]$ and $h_b[n]$, using **g dfilspec** and sketch the plots. Record the approximate stopband cutoff frequencies ω_{s1} and ω_{s2} that border the passband regions in $h_a[n]$ and $h_b[n]$.

(b) Consider downsampling these bandpass filters by a factor of 4. Using the frequency domain relationships in (4.13) that describe downsampling, sketch the magnitude responses for both bandpass filters assuming each is downsampled by a factor of 4. Now use **f dnsample** and **g dtft** to verify the accuracy of your sketch. Observe that the passband of the bandpass filter (which now occupies the region $-\pi < \omega < \pi$) is corrupted for the second filter because the integer band sampling condition has been violated.

(c) The difficulty encountered in part (b) clearly represents a problem for multirate implementations of bandpass filters that do not satisfy the integer band sampling condition. In such cases, we can often get around this difficulty by choosing a smaller integer downsampling factor, M, such that

$$M \leq \frac{\pi}{\omega_{s2} - \omega_{s1}}. \tag{4.24}$$

In addition, M must satisfy the conditions

$$\omega_{s1} \geq \pi k/M \tag{4.25}$$

$$\omega_{s2} \geq (k+1)\pi/M \tag{4.26}$$

for some integer value k between 0 and $M - 1$. What we are doing is reducing the decimation rate until the passband of the bandpass filter lies within an integer band. For narrow bandwidth filters, it is often necessary to check many values of M to find the best choice.

Determine the best value of M for a multirate implementation of $h_b[n]$. What is the computational gain ratio associated with this multirate filter implementation? □

EXERCISE 4.2.18. **Multirate Highpass Filters**

The implementation of multirate highpass filters is similar to that for lowpass and bandpass filters with only minor modifications. If we let the system function for the highpass filter, $H_{hp}(z)$, assume the form

$$H_{hp}(z) = 1 - H_{lp}(z)$$

(where $H_{lp}(z)$ is a lowpass filter) and use a multirate implementation to realize $H_{lp}(z)$, we have a multirate implementation for a highpass filter. A block diagram of this structure is given in Figure 4.8, where, in this case, the frequency selective filter, $H_{FS}(z)$, is a lowpass filter.

Consider the design of a multirate highpass filter with $\omega_c = \pi/4$. First design a zero-phase FIR lowpass filter with length 49 using **f fdesign** with the *Hamming window* option. Use **f lshift** to shift the filter by -24 to make it zero phase and use this zero-phase lowpass filter for $H(z)$ and $G(z)$ in the multirate implementation of $H_{lp}(z)$.

(a) Write a macro that implements the multirate highpass filter following the block diagram in Figure 4.8. For simplicity, use the nonpolyphase form to implement the decimator and interpolator. This can be done in only a few steps using **f convolve, f dnsample, f upsample,** and **f subtract**. Generate an impulse using **f siggen** and use it as the input to the macro. Compute the impulse response, $h[n]$, of the multirate filter. Display its magnitude response using **g dtft** and sketch the plot.

(b) Would the multirate highpass implementation work if the decimation and interpolation filters $H(z)$ and $G(z)$ were not zero-phase filters, but instead were linear phase filters with starting points at $n = 0$? Explain. Would the implementation work if $H(z)$ and $G(z)$ were of even length? Explain.   □

EXERCISE 4.2.19. **Multirate Frequency Selective Filters**

From the previous discussions and exercises, it should be clear that a variety of frequency selective filters can be designed as multirate filters using the structures shown in Figures 4.7 and 4.8.

(a) Consider the design of a bandstop filter. What type of filter (lowpass, highpass, bandpass) should be used in the multirate structure shown in Figure 4.8?

(b) How would you design an efficient multirate lowpass filter with a cutoff frequency of $7\pi/8$?

(c) How would you design a multirate Hilbert transformer for narrowband lowpass inputs?   □

REFERENCES

[1] P. A. Regalia, S. K. Mitra, and P. P. Vaidyanathan, "The Digital All-Pass Filter: A Versatile Signal Processing Building Block," *Proc. IEEE*, Vol. 76, No. 1 (Jan. 1988), pp. 19–37.

[2] P. P. Vaidyanathan, S. K. Mitra, and Y. Neuvo, "A New Approach to the Realization of Low Sensitivity IIR Digital Filters," *IEEE Trans. on Acoust., Speech, and Signal Proc.*, Vol. 26 (Apr. 1986), pp. 350–361.

[3] R. Ansari and B. Liu, "A Class of Low Noise Computationally Efficient Recursive Digital Filters with Applications to Sampling Rate Alterations," *IEEE Trans. on Acoust., Speech, and Signal Proc.*, Vol. 33 (Feb. 1986), pp. 90–97.

[4] D. Esteban and C. Galand, "Application of Quadrature Mirror Filters to Split-band Voice Coding Schemes," *Proc. 1977 IEEE Int. Conf. Acoustics, Speech, Signal Processing*, pp. 191–195.

[5] T. P. Barnwell III, "Sub-band Coder Design Incorporating Recursive Quadrature Filters and Optimum ADPCM Coders," *IEEE Trans. on Acoust., Speech, and Signal Processing*, Vol. 30, (Oct. 1982), pp. 751–765.

[6] R. Crochiere and L. Rabiner, *Multirate Digital Signal Processing*, Prentice-Hall, Inc., 1983.

[7] P. P. Vaidyanathan, *Multirate Signal Processing and Filter Banks,* Prentice-Hall, Inc., 1993.

[8] J. W. Woods (ed.) *Subband Image Coding*, Kluwer Academic Publishers, 1991.

[9] A. Ankansu, R. Haddad, and H. Caglar, "Perfect Reconstruction Binomial QMF-Wavelet Transform," *1990 SPIE Conf. Visual Commun. Image Processing*, pp. 609–618.

[10] M. Smith and T. Barnwell, "A Procedure for Designing Exact Reconstruction Filter Banks for Tree-Structured Subband Coders," *Proc. 1984 IEEE Int. Conf. on Acoustics, Speech, and Signal Proc.*, pp. 27.1.1–27.1.4.

[11] M. Bellanger, J. Daguet, and G. Lepagnol, "Interpolation, Extrapolation, and Reduction of Computational Speed in Digital Filters," *IEEE Trans. on Acoust., Speech, and Signal Proc.*, Vol. 22 (Aug. 1974), pp. 231–235.

[12] I. Daubechies, "Orthonormal Bases of Compactly Supported Wavelets," *Comm. on Pure and Applied Math.*, Vol. 51 (1988), pp. 909–996, 1988.

[13] L. Gazsi, "Explicit Formulas for Lattice Wave Digital Filters," *IEEE Trans. on Circuits and Systems*, Vol. 32 (January 1985), pp. 68–88.

Filter Structures

5

Until now this text has considered the system function with its specified region of convergence to be a complete description of an LTI digital filter. This assumption is true in the sense that each filter can be uniquely represented by one and only one system function and its associated ROC.[1] However, there are generally many ways in which a digital filter can be implemented. Some of these offer distinct advantages over others. Some implementations require a minimum number of multiplications for high-speed computation; others require fewer delays for minimum storage. Still other filters can be implemented in modular and partitioned forms for efficient VLSI circuit implementation, or in structures that are insensitive to quantization effects.

One of the primary reasons for studying filter structures is that fixed-point DSP processors are often used to implement filters. When finite precision arithmetic is used to realize a digital filter, nonideal behavior can, and usually will, occur. This happens for a variety of reasons: the filter coefficients must be quantized; the results of multiplications must be rounded or truncated; and fixed-point registers can overflow after additions. All of these limitations introduce distortions into the processing. Alternative realizations of digital filters are the most common means for dealing with these problems. Flow graphs, which were discussed in the companion text,[2] provide a convenient way to describe these various filter implementations. The following section presents a brief review of flow graphs. Some basic structures are then examined in terms of their implementation, delay, and computational efficiency. In the last part of the chapter, various finite precision effects are discussed and different structures are compared in this context.

---

[1] Remember that a system function by itself may represent several filters, each with a different ROC or region of convergence. But specification of *both* the ROC and the system function uniquely defines the filter.

[2] M. J. T. Smith and R. M. Mersereau, *Introduction to Digital Signal Processing: A Computer Laboratory Textbook,* John Wiley: New York, 1992.

5.1 FLOW GRAPHS AND BASIC FILTER STRUCTURES

Flow graphs can be used to describe a wide variety of systems. They offer a graphic summary of the computations involved and they are closely related to the block diagram representations that are typically found in introductory texts on systems theory. Unlike block diagrams, which sometimes lump several operations into a single block, flow graphs represent systems at a very basic level. For linear, time-invariant (LTI) filters, only a small set of operations—constant gains, branches, adders, and delays—are needed; these operators are precisely the ones contained in the difference equations. You should have had experience in representing LTI systems (or filters) as system functions, as difference equations, as pole/zero plots, and as flow graphs. Therefore, we will not discuss the mechanics of conversion among these various representations, but will instead focus on standard flow graphs for digital filters.

The starting point for the discussion is a causal IIR filter of the form

$$H(z) = \frac{\sum_{\ell=0}^{M} b_\ell z^{-\ell}}{1 + \sum_{\ell=1}^{N} a_\ell z^{-\ell}}, \tag{5.1}$$

which can be implemented using the difference equation

$$
\begin{aligned}
y[n] &= b_0 x[n] + b_1 x[n-1] + b_2 x[n-2] + \cdots + b_M x[n-M] \\
&\quad - a_1 y[n-1] - a_2 y[n-2] - \cdots - a_N y[n-N].
\end{aligned}
\tag{5.2}
$$

The equation involves the sum of $M+1$ input terms, $x[n]$, $x[n-1]$, ..., $x[n-M]$; and N output terms, $y[n-1]$, ..., $y[n-N]$, each of which is weighted by a filter coefficient. We can group the weighted input terms together as a single difference equation; the output terms can be similarly grouped. This yields the pair of coupled difference equations:

$$w[n] = b_0 x[n] + b_1 x[n-1] + b_2 x[n-2] + \cdots + b_M x[n-M] \tag{5.3}$$

$$y[n] = -a_1 y[n-1] - a_2 y[n-2] - \cdots - a_N y[n-N] + w[n]. \tag{5.4}$$

These coupled equations can be viewed as a cascade of two subsystems, the first having a system function that is the numerator of 5.1 and the second having a system function that is 1 divided by the denominator of 5.1. This implementation implied by this decomposition is called a *direct form I* realization. Its flow graph for the special case $M = N = 4$ is shown in Figure 5.1.

Two common parameters that can be used to compare structures are the number of delay elements and the number of multiplications that must be computed to evaluate each sample of the output sequence. Notice that in the direct form I structure there are $M + N + 1$ terms $x[n]$, $x[n-1]$, $x[n-2]$, ..., $x[n-M]$, $y[n-1]$, $y[n-2]$, ..., $y[n-N]$, each of which is multiplied by a coefficient and summed. The number of multiplications to be computed for each output sample is equal to the number of

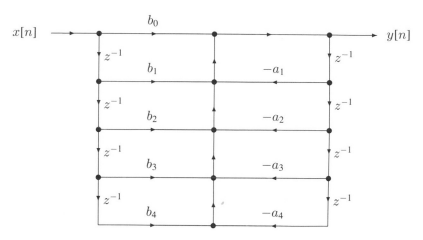

Figure 5.1. Direct form I realization of a fourth-order filter.

nontrivial filter coefficients.[3]

We derived the direct form I structure by writing one difference equation for the numerator of (5.1) and another for the denominator. Each difference equation corresponds to a subsystem. The first subsystem has as its input the sequence $x[n]$ and as its output the sequence $w[n]$; the second has input $w[n]$ and output $y[n]$. If the order of these two subsystems is reversed, many of the stored samples in the first subsection (the denominator part) can be shared with the second subsection (the numerator part). This is the *direct form II* or *canonic direct form* realization. The direct form II realization has the advantage that it requires the minimum number of delays ($\max[M, N]$) and the minimum number of multiplications ($M + N + 1$) among the classical implementation structures.

The difference equations that describe the direct form II implementation of $H(z)$ are

$$w[n] = x[n] - a_1 w[n-1] - a_2 w[n-2] - \cdots - a_N w[n-N] \qquad (5.5)$$

$$y[n] = b_0 w[n] + b_1 w[n-1] + b_2 w[n-2] + \cdots + b_M w[n-M]. \qquad (5.6)$$

The flow graph for a direct form II realization of a fourth-order filter is shown in Figure 5.2. Notice the reduced number of delay elements. The quantities being stored are not past values of the input and output sequences, but instead are past values of the intermediate sequence $w[n]$.

Filters may also be implemented as a cascade of first- and second-order subsystems (or sections). This is achieved by factoring the numerator and denominator polynomials of the system function to put it into the form

$$H(z) = A \prod_{m=1}^{I} \frac{b_{0,m} + b_{1,m} z^{-1} + b_{2,m} z^{-2}}{1 + a_{1,m} z^{-1} + a_{2,m} z^{-2}} \qquad (5.7)$$

---

[3]Multiplications by the constants 1, -1, and 0 are typically not included when counting multiplications.

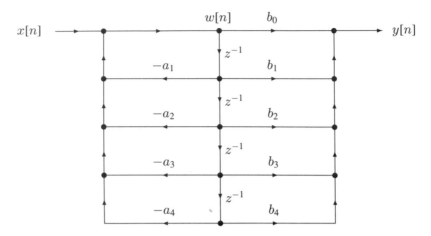

Figure 5.2.   Direct form II realization of a fourth-order filter.

and then realizing each second-order filter with one of the direct form structures. This *cascade* realization is perhaps the most common structure for implementing recursive digital filters. A large number of cascade structures are possible since there are many ways that the numerator factors can be paired with the denominator factors, many possible orderings of the sections, and several possible realizations for each section.

Since the direct form II structure has the minimum number of delays and multi-plications for a general Nth-order filter, we will realize our cascade implementations using direct form II second-order sections. This cascade realization corresponds to the difference equations:

$$w_1[n] = x[n] - a_{1,1}w_1[n-1] - a_{2,1}w_1[n-2]$$
$$w_2[n] = b_{0,1}w_1[n] + b_{1,1}w_1[n-1] + b_{2,1}w_1[n-2] \tag{5.8}$$

$$w_3[n] = w_2[n] - a_{1,2}w_3[n-1] - a_{2,2}w_3[n-2]$$
$$w_4[n] = b_{0,2}w_3[n] + b_{1,2}w_3[n-1] + b_{2,2}w_3[n-2] \tag{5.9}$$

$$\vdots$$

$$w_{2I-1}[n] = w_{2I-2}[n] - a_{1,I}w_{2I-1}[n-1] - a_{2,I}w_{2I-1}[n-2]. \tag{5.10}$$
$$y[n] = b_{0,I}w_{2I-1}[n] + b_{1,I}w_{2I-1}[n-1] + b_{2,I}w_{2I-1}[n-2]. \tag{5.11}$$

Expressing the general filter of (5.1) as a sum of partial fractions results in

$$H(z) = C + \sum_{\ell=1}^{I} \frac{b_{0,\ell} + b_{1,\ell}z^{-1}}{1 + a_{1,\ell}z^{-1} + a_{2,\ell}z^{-2}}, \tag{5.12}$$

which leads to a *parallel form* implementation. If again we use the direct form II structure for each second-order section, the corresponding difference equations are

$$w_1[n] = x[n] - a_{1,1}w_1[n-1] - a_{2,1}w_1[n-2]$$

Table 5.1.   Coefficients For Elliptic Filter Implementations.

Direct Form	Cascade	Parallel
$b_0 = 0.0802$	$b_{0,1} = 0.1658$	$b_{0,1} = -0.2646$
$b_1 = 0.1935$	$b_{1,1} = 0.2777$	$b_{1,1} = 0.7267$
$b_2 = 0.2595$	$b_{2,1} = 0.1658$	$a_{1,1} = -0.5947$
$b_3 = 0.1935$	$a_{1,1} = -0.5947$	$a_{2,1} = 0.2042$
$b_4 = 0.0802$	$a_{2,1} = 0.2042$	$b_{0,2} = -0.2102$
$a_1 = -0.9666$	$b_{0,2} = 0.4835$	$b_{1,2} = -0.2202$
$a_2 = 1.1330$	$b_{1,2} = 0.3567$	$a_{1,2} = -0.3719$
$a_3 = -0.4968$	$b_{2,2} = 0.4835$	$a_{2,2} = 0.7076$
$a_4 = 0.1445$	$a_{1,2} = -0.3719$	$C = 0.555$
	$a_{2,2} = 0.7076$	

(Coefficients for direct form implementation are stored in the file **f010**. For the cascade and parallel forms, the coefficients are stored in the files **f010c1**, **f010c2**, and **f010p1**, **f010p2**, **f010p3**, respectively.)

$$v_1[n] = b_{0,1}w_1[n] + b_{1,1}w_1[n-1] \qquad (5.13)$$

$$w_2[n] = x[n] - a_{1,2}w_2[n-1] - a_{2,2}w_2[n-2]$$
$$v_2[n] = b_{0,2}w_2[n] + b_{1,2}w_2[n-1] \qquad (5.14)$$

$$\vdots$$

$$w_I[n] = x[n] - a_{1,I}w_I[n-1] - a_{2,I}w_M[n-I]$$
$$v_I[n] = b_{0,I}w_I[n] + b_{1,I}w_I[n-1] \qquad (5.15)$$

$$y[n] = C\delta[n] + v_1[n] + v_2[n] + \cdots + v_I[n]. \qquad (5.16)$$

Any filter that can be expressed in the form of (5.1) can be implemented as a direct form I, direct form II, cascade of second-order sections, or parallel form realization (providing $M = N$). To illustrate this point, Table 5.1 is included, which gives the coefficients for each of these implementations for a fourth-order elliptic lowpass filter. When the denominator coefficients $a_1, a_2, \ldots, a_N$ are zero, the filter is an FIR filter. FIR filters are a special case of IIR filters.

Lattice structures constitute another useful class of implementations. They have been researched extensively for autoregressive modeling. An extensive treatment of the history and motivation for these structures is beyond the scope of this book. Instead, we will consider lattices strictly as implementations for realizing digital filters. They are very different from the direct, cascade, and parallel forms and consequently provide an interesting contrast. Figure 5.3 shows an AR-lattice (or Auto-Regressive lattice) structure that can be used to realize filters that have only poles (i.e., no zeros). Such filters are typically called *allpole* or *auto-regressive* filters. The coefficients, k_i, are known as *PARCOR* (*partial correlation*) or *reflection coefficients*. We see that this structure is a cascade of two-ports. Notice that the first stage shown in the lattice is actually simpler

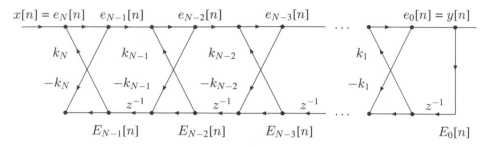

$$x[n] = e_N[n] \quad e_{N-1}[n] \qquad e_{N-2}[n] \qquad e_{N-3}[n] \qquad \cdots \qquad e_0[n] = y[n]$$

Figure 5.3. Flow graph of an AR-lattice.

than indicated in the figure because the branch with the $-k_N$ coefficient multiplier can be removed since its output node is not used. This branch is only included to show the modularity of the lattice. (If this node is retained and used as the output node of the filter, the resulting filter is an allpass.) The lattice coefficients may be related to the coefficients of the direct form allpole filter

$$\frac{1}{A(z)} = \frac{1}{1 + \displaystyle\sum_{m=1}^{N} a_m z^{-m}}$$

by the following recursion, which we will call the Levinson recursion:[4]

1. for $m = 1$ to N **3.** $k_1 = a_1^{(1)}$.

$\qquad a_m^{(N)} = a_m.$

2. for $i = N$ to 2

$\qquad$ a. $k_i = a_i^{(i)}$

$\qquad$ b. for $m = 1$ to $i - 1$

$$a_m^{i-1} = \frac{a_m^{(i)} - k_i a_{i-m}^{(i)}}{1 - k_i^2}.$$

In this recursion, you will notice the terms $a_m^{(i)}$. These are the direct form coefficients of the allpole filters that describe the system functions of shortened versions of the lattices. They are useful for system modeling, but for our purposes they can simply be discarded when the recursion is completed.

To illustrate the recursion, consider the allpole filter formed from the denominator of the elliptic filter in Table 5.1.

$$\frac{1}{A(z)} = \frac{1}{1 - 0.9666z^{-1} + 1.1330z^{-2} - 0.4968z^{-3} + 0.1445z^{-4}} \qquad (5.17)$$

---

[4]This recursion often has the names Levinson and/or Durbin associated with it. This form of the equations is referred to as the *backward recursion*.

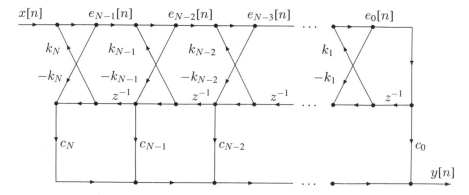

Figure 5.4. A lattice structure for filters with both poles and zeros.

The initial coefficient assignments are $a_1^{(4)} = -0.9666$, $a_2^{(4)} = 1.133$, $a_3^{(4)} = -0.4968$, $a_4^{(4)} = 0.1445$. Beginning with $i = 4$, we proceed to calculate $k_4 = a_4^{(4)} = a_4 = 0.1445$. Then we solve for $a_m^{(3)}$ for $m = 1, 2, 3$ and repeat the process until all four reflection coefficients have been computed. The computation of the lattice coefficients for this specific case proceeds as follows:

$$k_4 = 0.1445,$$

$$a_1^{(3)} = \frac{a_1^{(4)} - k_4 a_3^{(4)}}{1 - (k_4)^2} = -0.9138947,$$

$$a_2^{(3)} = \frac{a_2^{(4)} - k_4 a_2^{(4)}}{1 - (k_4)^2} = -0.9899519,$$

$$a_3^{(3)} = \frac{a_3^{(4)} - k_4 a_1^{(4)}}{1 - (k_4)^2} = -0.3647422,$$

$$k_3 = a_3^{(3)} = -0.3647422,$$

$$a_1^{(2)} = \cdots$$

$$\vdots$$

$$k_1 = a_1^{(1)} = -0.3628413. \tag{5.18}$$

The reflection coefficients are

$$k_1 = -0.36284,$$
$$k_2 = 0.75737,$$
$$k_3 = -0.36474,$$
$$k_4 = 0.1445. \tag{5.19}$$

Figure 5.4 shows a generalization of the lattice structure [1,2] that can implement filters with both poles and zeros. These are sometimes called *Auto-Regressive Moving*

Average or (*ARMA*) filters. ARMA filters have system functions that are similar to (5.1). The numerator of the system function is created by the coefficients $\{c_m\}$. These can be found from the direct form coefficients b_i and a_i, by first using the Levinson recursion to find the reflection coefficients k_i. Then the additional coefficients, c_m, can be computed recursively using the relation

$$c_m = b_m - \sum_{i=m+1}^{N} c_i a_{i-m}^{(i)} \quad m = N - 1, \ldots, 0 \tag{5.20}$$

with $c_N = b_N$. For the elliptic filter in (5.1), the $\{k_i\}$ were given in (5.19). The additional parameters resulting from (5.20) are

$$
\begin{aligned}
c_4 &= 0.0802, \\
c_3 &= 0.27102, \\
c_2 &= 0.41638, \\
c_1 &= 0.2305, \\
c_0 &= -0.0642.
\end{aligned}
\tag{5.21}
$$

Like the other structures we have seen, the general lattice structure corresponds to a set of difference equations. Initially set

$$
\begin{aligned}
e_N[n] &= x[n], \\
f_{N+1} &= 0.
\end{aligned}
\tag{5.22}
$$

Then for $i = N, N - 1, \ldots, 1$ set

$$e_{i-1}[n] = e_i[n] + k_i E_{i-1}[n - 1], \tag{5.23}$$

$$E_i[n] = E_{i-1}[n - 1] - k_i e_{i-1}[n], \tag{5.24}$$

$$f_i[n] = f_{i+1}[n] + c_i E_i[n]. \tag{5.25}$$

Finally,

$$y[n] = f_1[n] + c_0 e_0[n]. \tag{5.26}$$

There are many other structures that can be used to implement digital filters. Included in these are variations that can be found by using the transposition theorem. (This theorem states that the system function for a flow graph and its transpose are the same. The transpose of a network is formed by first interchanging the input and output nodes and then reversing the directions of the arrows along each branch in the flow graph.) Thus, for every structure that we propose, we also have a corresponding transposed structure. A particularly important realization that can be used for many important filters, including those designed using the techniques described in Chapter 3, is the realization of the system as a parallel connection of two allpass filters that was discussed in Chapter 4. This realization requires fewer multiplications than the realizations discussed in this chapter.

5.2 REALIZING STRUCTURES

One important issue that we must confront is the fact that hardware realizations of digital filters require that the coefficients be quantized and that products be rounded or truncated. These nonlinear operations introduce errors into the system outputs as discussed in the next two sections.

To understand the inner workings of IIR filter structures better, the functions **f direct1** and **f direct2** will be used. They evaluate the difference equations for the direct form I and II structures using both fixed-point (i.e., quantized) arithmetic and floating point arithmetic (which closely approximates the unquantized case). These functions receive as their input a file whose contents are then filtered, and written to a file. The functions are unique in that they have a mode of operation in which the multiplication operations within the difference equation can be displayed on the screen. This allows the effects of the quantization of products to be seen for each output sample. By using these functions, the impact of coefficient quantization and roundoff or truncation errors can be examined at each node in the flow graph. This provides a convenient method for studying these effects.

In our examination of FIR filters, macros will be used to build implementations for different structures. These macros will be used to realize the specific difference equations that define the structure. As an example, consider the length 9 FIR filter

$$h[n] = 0.1\delta[n] + 0.2\delta[n-1] + 0.3\delta[n-2] + \cdots + 0.8\delta[n-7].$$

The functions **f lshift**, **f gain**, and **f add** can be used to delay $x[n]$ appropriately, multiply the appropriate samples by the proper coefficients, and sum the products together. The direct form realization[5] of this filter is

$$y[n] = 0.1x[n] + 0.2x[n-1] + \cdots + 0.8x[n-7].$$

It can be implemented using the macro **df.bat** given in Table 5.2.[6] The macro is invoked by typing

df   xn   yn

where **xn** is the name of the input file and **yn** is the name of the output file. In the examples that follow, you will examine structures for both IIR and FIR filters using these tools.

EXERCISE 5.2.1. **A Review of Basic Structures**

Direct form, cascade, and parallel structures are discussed in the companion text *Introduction to Digital Signal Processing: A Computer Laboratory Textbook*. This exercise focuses on simple conversions among these implementations and is intended as a warm-up. It does not involve the use of a computer. Consider the simple causal second-order

---

[5]Note, for FIR filters the direct form I and II implementations are the same.

[6]Parentheses are used in place of brackets in the macro because DOS does not recognize brackets as valid characters in a file name.

Table 5.2. The **df.bat** Macro for Implementing a Direct Form FIR Filter.

f gain	%1	x0	0.1
f lshift	%1	x(n-1)	1
f gain	x(n-1)	x1	0.2
f lshift	%1	x(n-2)	2
f gain	x(n-2)	x2	0.3
f lshift	%1	x(n-3)	3
f gain	x(n-3)	x3	0.4
f lshift	%1	x(n-4)	4
f gain	x(n-4)	x4	0.5
f lshift	%1	x(n-5)	5
f gain	x(n-5)	x5	0.6
f lshift	%1	x(n-6)	6
f gain	x(n-6)	x6	0.7
f lshift	%1	x(n-7)	7
f gain	x(n-7)	x7	0.8
f add	x0	x1	store
f add	store	x2	store
f add	store	x3	store
f add	store	x4	store
f add	store	x5	store
f add	store	x6	store
f add	store	x7	%2

filter,

$$H(z) = \frac{0.1 + 0.2z^{-1} + 0.1z^{-2}}{1 - 0.75z^{-1} + 0.125z^{-2}}. \tag{5.27}$$

(a) Sketch the flow graphs corresponding to the direct form I and direct form II implementations of this filter.

(b) Sketch the flow graph depicting a cascade of first-order sections.

(c) Sketch the flow graph depicting a parallel form implementation using first-order sections.

(d) Redraw the flow graphs of parts (a–c) in their transposed forms.   □

EXERCISE 5.2.2. Implementing Basic Filter Structures

Use **f rgen** to create a 12-point zero mean random sequence, $x[n]$ with its starting point at $n = 0$ using a seed value of 1.

(a) Use **g dtft** to display and sketch the DTFT magnitude of $x[n]$.

(b) Consider the direct form I implementation of the elliptic filter $H(z)$, whose coefficients are given in Table 5.1. This filter is also provided for you in the file **f010**. Display and sketch the magnitude response using **g dtft**. Write the difference equations that correspond to the direct form I implementation [following (5.3) and (5.4)].

Apply the function **f direct1** to evaluate the output using $x[n]$ as the input and $H(z)$ as the filter. Do this by typing **f direct1**, pressing the "enter" key, and responding to the program prompts that follow. Use files **out1** and **out2** for the unquantized and quantized output files, -1 and 1 as the minimum and maximum quantization values, and 100 for the number of levels. Select options 0 for saturation arithmetic, and 0 for the verbose mode of operation. The function will display the intermediate multiplications and quantized products involved in implementing the difference equation for $n = 0$. Pressing the "c" key allows you to manually step through the difference equation for successive values of n. Note that the quantized output $\hat{y}[n]$, which appears as $qy[\cdot]$ on the screen, is obtained by adding the $b_i \times x[n-i]$ terms and subtracting the $a_i \times y[n-i]$ terms. In other words, it is a direct implementation of (5.2) with quantization. Since the number of levels is fairly large in this case, there should be good agreement between the quantized and unquantized values.

Next, evaluate 50 samples of the output by running **f direct1** in its nondisplay (i.e., nonverbose) mode. Display and sketch the DTFT magnitude of **out1** and **out2** using **g dtft**. Notice that the DTFT of **out1**, which represents the results of floating point processing, is a lowpass filtered version of $x[n]$. However, the DTFT of **out2** is very distorted. This is due to the quantization and roundoff error in the filtering process.

(c) Repeat part (b) for the case of a direct form II implementation using the function **f direct2**. Notice that distortion is also present in the fixed point output but that the distortion is different from that in the direct form I case.

(d) A cascade realization can be implemented using a direct form II structure for the second-order sections. Sketch the flow graph for the elliptic filter implemented as a cascade of second-order direct form II sections. Include the coefficients for the filter, which are given in Table 5.1, in your flow graph. Write the difference equations [following (5.8–5.11)] that describe the cascade structure for this filter.

(e) A parallel realization can also be implemented. Sketch the flow graph corresponding to a parallel form realization based on second-order direct form II sections. Again include the coefficient values in your flow graph, all of which may be found in Table 5.1. Write the difference equations [following (5.13–5.15)] for the parallel implementation.   □

EXERCISE 5.2.3.   **A Preview of Sensitivity Issues**

Read Exercise 5.2.2 if you have not already done so. The same input and fourth-order elliptic filter are used again in this problem.

Processors for implementing digital filters rely on quantized number representations and they have a limited dynamic range. If intermediate values in a calculation exceed that range, errors are introduced into the system output. This issue is a topic of further discussion in the subsequent sections. In general, the intermediate sample values in well-chosen structures remain confined to a relatively small amplitude range.

Here you will consider direct form I, and cascade and parallel implementations that use direct form II second-order sections. Create a 12-point random input sequence using

f rgen with a seed value of 1. Filter this input with the elliptic filter in file **f010** and measure the peak signal amplitude (i.e., the maximum absolute value of $w[n]$) that occurs inside each section using **f direct2** with minimum and maximum quantization values of -1 and $+1$. Select 1000 levels, 0 for saturation arithmetic, and 0 for the verbose operation mode.

The quantized product values for $w[n]$ appear in the second column of numbers. The sum of these terms is shown as $qw[\cdot]$. Calculate twelve output samples and record their peak magnitude value. For the direct form realizations, the filter coefficients are stored in the file **f010**. For the cascade and parallel implementations, the filter coefficients for the different sections are stored in the files **f010c1**, **f010c2**, and **f010p1**, **f010p2**, **f010p3**, respectively. Measure the peak values within each section using the input $x[n]$. Based on this examination, determine which structure might be most likely to saturate when finite precision arithmetic is used. We will see the impact of structural sensitivity in later exercises. □

EXERCISE 5.2.4.   **Sketching the Frequency Response of a Lattice Filter**

Figure 5.4 depicts a generalized lattice structure for implementing filters with poles and zeros. Assume that the filter is of second order with lattice coefficients

$$k_1 = 0.9055127 \qquad k_2 = -0.591028$$

$$c_0 = 0.4287955 \qquad c_1 = 0.7262529 \qquad c_2 = 0.46631.$$

Examine the key nodes in the flow graph and write a difference equation at each node based on the branches entering and leaving the node. From these difference equations, calculate the system function for the network. Sketch the direct form II flow graph. Use the *create file* option in **f siggen** to create a file containing this filter, and display and sketch its DTFT magnitude. □

EXERCISE 5.2.5.   **Nocomputable Networks**

While flow graphs provide a layout for implementing a digital filter, not all flow graphs are implementable. An implementable realization requires the ability to compute all the network node variables when they are needed. This is not always possible, as illustrated in Figure 5.5. Observe that the difference equations cannot be solved successively unlike the other sets of difference equations considered previously. Flow graphs of this type are called *noncomputable* flow graphs.

The pair of difference equations that describe this network is

$$w[n] = x[n-1] + 1/2\,w[n-1] + y[n],$$

$$y[n] = x[n] + 1/3\,w[n].$$

Observe that neither one of these equations can be evaluated independently for a particular value of n. In the first case, $y[n]$ is unavailable, whereas in the second, $w[n]$ is unavailable.

Modify the difference equations so that they can be solved recursively. Sketch the new flow graph corresponding to the modified equations. □

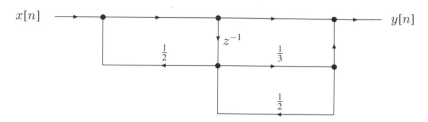

Figure 5.5.   A noncomputable flow graph.

EXERCISE 5.2.6.   **Simple FIR Filter Implementation**

FIR filters are special cases of IIR filters formed when the denominator polynomial coefficients $a_1, a_2, \ldots, a_N$ of the IIR filter system function are zero. Based on the general $(N-1)$th-order direct form II network, sketch the flow graph for a length-N FIR filter. Redraw the graph so that it resembles a ladder oriented horizontally. FIR networks of this type are sometimes called *tapped delay lines* or *transversal filter structures*.

Design an eight-point lowpass filter, $h[n]$, with a cutoff frequency of $\pi/3 = 1.0472$, using the *Hamming window* option in **f fdesign**. You can view the filter coefficients by typing the file to the screen. Refer to the macro **df.bat** shown in Table 5.1. It is provided for you in the software. Copy it into another file, **df1.bat** and modify it using a text editor so that it implements your filter. Compute the impulse response of the filter using your macro **df1.bat** with **impulse** as the input. Truncate the output using **f truncate** to 15 samples. Record these output sample values. If the output values and filter impulse response are the same, your macro is working properly. Display and sketch the DTFT magnitude of this output.   □

EXERCISE 5.2.7.   **Linear-Phase FIR Filter Implementations**

We saw in Chapter 3 that linear-phase filters have symmetric impulse responses. Consider the linear phase filter, $h[n]$, defined as

$$h[n] = \delta[n] + 2\delta[n-1] + 3\delta[n-2] + 4\delta[n-3] + 3\delta[n-4]$$
$$+2\delta[n-5] + \delta[n-6]. \tag{5.28}$$

It can be implemented efficiently using a flow graph like the one shown in Figure 5.6 for $M = 6$. This implementation exploits the symmetry in the filter by pairing and adding input samples that will be multiplied by the same coefficient prior to that multiplication to reduce the number of multiplication operations that will have to be performed by the hardware.

(a) How many multiplications, additionss, and delays are required to implement the filter in (5.28) if a standard direct form II structure is used? Repeat the tally of multiplications, additions, and delays if the linear-phase structure of Figure 5.6 is used.

(b) Write the difference equations for the efficient linear phase structure in Figure 5.6. Following the discussion in Section 5.2 on macros for implementing FIR filters, write

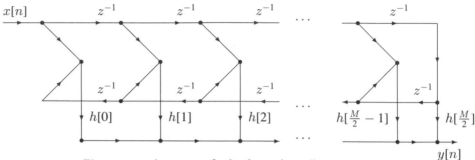

Figure 5.6. A structure for implementing a linear phase filter.

a macro that realizes the linear-phase filter in (5.28) based on Figure 5.6. Verify its proper operation by computing its impulse response. Do this by using the file **impulse** as the input to the macro.

(c) Now consider an efficient linear-phase flow graph for an even length filter. Determine and sketch the linear-phase flow graph that exploits coefficient symmetry for the filter

$$h[n] = \delta[n] + 2\delta[n-1] + 3\delta[n-2] + 4\delta[n-3]$$
$$+ 4\delta[n-4] + 3\delta[n-5] + 2\delta[n-6] + \delta[n-7]. \tag{5.29}$$

How many multiplications are required to implement the filter based on this flow graph? Write a macro that realizes the flow graph. Evaluate its response to a unit impulse (which is stored in the file **impulse**) and record the output.

Be aware that executing an implementation in the form of a macro involves reading and writing files to and from the hard disk. This overhead makes macros very slow to execute. Thus even though your linear-phase structure is efficient, a macro implementation cannot be expected to be very fast.   □

EXERCISE 5.2.8.   Cascade Structures for Linear-Phase FIR Filters

Linear-phase FIR filters can also be implemented in cascade structures. These are particularly efficient when the filters in the cascade are themselves linear phase.

Design a length 16 lowpass filter using the *rectangular window* option in **f fdesign**, with a cutoff frequency of $3\pi/4 = 2.3562$ rad. Display and sketch the pole/zero plot using **g polezero**. Adjust the unit circle size so that all the roots are displayed in the plot. Explicitly examine and record the zeros of the filter using the *list pole/zero* option. Observe that the roots occur either on the unit circle or in reciprocal root pairs off the unit circle. Moreover, every complex root has a complex conjugate root associated with it. In general, linear-phase filters are comprised of three categories of z-plane roots: roots occurring in complex conjugate reciprocal quadruples, which have the polynomial form $(1 + b_1 z^{-1} + b_2 z^{-2} + b_1 z^{-3} + z^{-4})$; roots occurring in complex conjugate or real reciprocal pairs, which have the form $(1 + b_1 z^{-1} + z^{-2})$; and the roots $1 - z^{-1}$ and $1 + z^{-1}$. Express the length 16 linear-phase filter as a cascade of these kinds of sections. The *delete root* and *write file* features in **g polezero** can be used to delete roots and write the various sections to output files. In this way you can create the cascade structure. Note, however, that the gain information will be lost. It will be necessary to rescale the filter so that its gain is unity in the passband. Assume that this gain term is applied at the output end of the flow graph.

Sketch an efficient flow graph, i.e., one that will implement the filter based on this kind of cascade using a minimum number of multiplications and include the new filter coefficients in the flow graph. How many multiplications are required to implement this filter in direct form, in the new cascaded form, and in the linear-phase form introduced in the previous exercise?   □

EXERCISE 5.2.9. Allpole Lattice Filters

The presence of zeros in the system function is generally a prerequisite for obtaining high-quality lowpass, highpass, and other frequency selective filters of reasonable order. Zeros enable the filter to have high attenuation in the stopband. In modeling applications where the filter is intended to represent the spectral shape of some processes and where regions of high attenuation are not present, allpole filters are sometimes useful. In this exercise, you will consider the implementation of an allpole filter in lattice form.

(a) Using the Levinson recursion compute the lattice coefficients corresponding to the filter

$$H(z) = \frac{1}{A(z)} = \frac{1}{1 - 0.5z^{-1} - 0.2z^{-2} - 0.1z^{-3}} \tag{5.30}$$

and sketch the lattice structure for this filter. You may find it convenient to compute the lattice coefficients by hand or by writing a small computer program to evaluate the recursion. Such a program can be run on a hand-held calculator.

Comment. The recursion used to generate the N lattice coefficients has a special property. In the process of performing the recursion, N sets of direct form coefficients are computed—one corresponding to a filter of first order, second order, third order, and so on. These are the direct form equivalents of the filters you would obtain if the sequence of lattice coefficients were truncated.

(b) Compare the number of multiplications and additions required to implement the IIR filter in part (a) using the direct form and lattice structures.

(c) (optional) Since the lattice structure is not composed of direct form components, it cannot be realized easily using **f direct2**. Write a program (in any language) that implements this lattice filter. Compute the impulse response for your lattice filter program and compare it with the impulse response of (5.30).   □

EXERCISE 5.2.10. FIR Lattice Filters

In modeling applications, FIR lattices are often used to implement inverse filters for allpole systems. The flow graph of a general length-N FIR lattice is shown in Figure 5.7.

When the AR-lattice coefficients, k_i, corresponding to the system function $H(z) = 1/A(z)$ are used in the FIR lattice structure, the resulting filter is an FIR filter with system function $A(z)$.

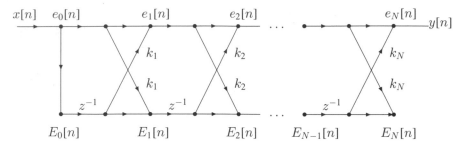

Figure 5.7. Flow graph of a general length-N FIR lattice.

Consider the FIR filter

$$H(z) = \sum_{m=0}^{7} b_m z^{-m}, \qquad (5.31)$$

where

$$
\begin{array}{llll}
b_0 = 1 & b_1 = 1.364033 & b_2 = 0.715342 & b_3 = -0.2377656 \\
b_4 = -0.35675 & b_5 = 0.092772 & b_6 = 0.25569 & b_7 = -0.18613.
\end{array}
$$

If we apply the Levinson recursion to this filter we obtain

$$
\begin{array}{llll}
k_1 = -0.7201 & k_2 = -0.759868 & k_3 = -0.762199 & k_4 = -0.56888 \\
k_5 = 0.74563 & k_6 = -0.527867 & k_7 = 0.18613.
\end{array}
$$

Because this is a seventh-order filter, computing the lattice coefficients by hand is more conveniently done using a program.

Write a macro that implements this filter in an FIR lattice structure. Using the file **impulse** as the input, compute and record the impulse response and verify that the macro works properly. Your output samples should equal $b_0, b_1, \ldots, b_7$. Construct a filter file using the *create file* option in **f siggen** that contains the filter $H(z)$. Compute and sketch its DTFT magnitude. $\square$

EXERCISE 5.2.11. General Lattice Structures

The chapter introduction presented recursive equations that could calculate the generalized lattice coefficients from the coefficients of the numerator and denominator polynomials in the system function.

Consider the causal filter with system function

$$H(z) = \frac{0.114177 - 0.175558z^{-1} + 0.114177z^{-2}}{1 + 1.005701z^{-1} + 0.4420148z^{-2}}. \qquad (5.32)$$

(a) Apply the recursion to calculate the lattice filter coefficients by hand or by using the computer. Use the *create file* option in **f siggen** to put $H(z)$ in a file. Use **f filter** with the file **impulse** as the input to compute the first 100 points of the filter's impulse response and sketch your results.

(b) (optional) Write a program in any language that is supported in your computing environment to implement the generalized lattice filter. Write the program so that it reads and writes files in the standard DSP format. Verify the program's proper operation by computing its impulse response using the method described above and compare the results with those found in part (a).   □

5.3 COEFFICIENT QUANTIZATION

Practical implementations of digital filters must use coefficients that can be represented with a limited number of bits. This implies that ideal coefficients must be either rounded or truncated to the register length of the machine after the filter is designed and before the filter is implemented. Coefficient quantization can lead to a noticeable distortion of the frequency response, the severity of which depends on the structure. This section examines the effects of coefficient quantization for the classical digital filter structures that the previous section introduced. Although detailed sensitivity analyses have been developed for studying coefficient quantization, such approaches are beyond the scope of this text and consequently are not treated here. Instead, several general rules of thumb are presented that have been shown to be effective in practice.

To begin the discussion, consider the impact of coefficient quantization on a direct form IIR[7] implementation of a system function

$$H(z) = \frac{\displaystyle\sum_{\ell=0}^{N} b_\ell z^{-\ell}}{1 + \displaystyle\sum_{\ell=1}^{M} a_\ell z^{-\ell}} = \frac{b_0 \displaystyle\prod_{m=1}^{N}(1 - \beta_m z^{-1})}{\displaystyle\prod_{m=1}^{M}(1 - \alpha_m z^{-1})}. \tag{5.33}$$

The numerator and denominator coefficients are b_ℓ and a_ℓ and the zeros and poles of the system lie at $z = \beta_m$ and $z = \alpha_m$, respectively. When the polynomial coefficents, $\{b_\ell, a_\ell\}$, are quantized, the new coefficients can be written as $\hat{b}_\ell = b_\ell + \Delta b_\ell$ and $\hat{a}_\ell = a_\ell + \Delta a_\ell$, where Δa_ℓ and Δb_ℓ denote coefficient errors. This error leads to non-linear changes in the locations of the zeros and poles. An important example illustrating the issue of sensitivity is the case for which the poles of $H(z)$ are real and distinct. The resulting changes in the denominator coefficients shift the poles by amounts $(\Delta\alpha_\ell)$ given by

$$\Delta\alpha_\ell = \sum_{m=1}^{M} \frac{\Delta a_m \alpha_\ell^{N-m}}{\displaystyle\prod_{j=1, \, j\neq\ell}^{M} (\alpha_\ell - \alpha_j)}. \tag{5.34}$$

From this we see that filters with poles that are bunched closely together are very sensitive to changes in the polynomial coefficients, because from (5.34) we see that roots in close

---

[7]Since the coefficients of direct form I and direct form II implementations are the same, we will treat these (and their transposes) together as direct form implementations.

proximity cause the denominator term to become very small. This, in turn, produces a large movement in the poles.

High order direct form realizations of recursive filters are sensitive to coefficient quantization. Cascade and parallel realizations based on first- and second-order sections tend to be much less sensitive. This fact, along with the fact that cascade and parallel realizations require approximately the same amount of memory and computation as the direct form realizations, is why these implementations are commonly used in practice.

The issue of structures is somewhat less important for FIR filters. The severity of coefficient sensitivity is not dramatic for FIR filters and coefficient quantization is usually less of a problem. In fact, direct form realizations are quite popular for FIR filters. The following exercises explore these and related issues.

EXERCISE 5.3.1.   Coefficient Sensitivity in Direct Form Implementations

The preceding discussion suggested that the sensitivity of the direct form implementations could be predicted from the pole/zero plot. To illustrate this, examine the direct form filters $H_0(z)$, $H_1(z)$, and $H_2(z)$ stored in the DSP files **h0df**, **h1df**, and **h2df**, respectively.

(a) Using **g polezero**, display and sketch the pole/zero plots and magnitude responses for each of these filters. Based on the pole/zero plots, predict which filters will be sensitive to coefficient quantization.

(b) Consider quantizing the coefficients of each filter to 2048 levels (which is equivalent to 11 bit quantization). As you examine the coefficients, you will observe that the dynamic ranges of the numerator and denominator coefficients are very different. If you blindly quantize the coefficients based on the maximum coefficient value, many of the numerator coefficients (because their amplitudes are smaller) will be disproportionately distorted. However, if you rescale the range of the numerator and denominator so that they are approximately equal, you will achieve a superior quantized coefficient representation. The same end can be achieved by separately quantizing the numerator and denominator polynomials.

Type the files containing each of the filters to the screen. Find the minimum and maximum values for both the numerator and denominator in each of the filters and use these values to specify the quantization range values. Remember that the leading coefficient in the denominator is 1, although it is not shown in the file. Thus the maximum value can never be less than 1 for the denominator. Quantize the filter coefficients by first using **f convert** to split the filters into numerator and denominator files, separately quantizing the files to 2048 levels using **f quantize**. Then merge the quantized numerator and denominator files back into a direct form file using **f revert**.

Use **g polezero** to display and sketch the pole/zero plots and magnitude responses of the quantized filters. Verify the accuracy of your prediction in part (b).

(c) Why do narrowband bandpass filters seem to be more sensitive to coefficient quantization than lowpass filters?   □

EXERCISE 5.3.2.   **Truncating and Rounding Filter Coefficients**

This exercise will examine the difference between truncating and rounding the filter coefficients of the bandpass filter, $H(z)$, which is provided in the file **h0df**. Copy this filter into another file **ht** to form $H_t(z)$. Using a text editor, manually *truncate* the filter coefficients so that they only include four digits after the decimal point. In another copy of the file, called **hr**, *round* the coefficients to four digits to produce the filter $H_r(z)$.

Using **g polezero**, display and sketch the pole/zero plots and magnitude responses for **ht** and **hr**. Do you observe a significant difference in the two pole/zero plots? What about the magnitude response plots? Using the file **impulse** as input, apply the function **f filter** and compute 60 points of the impulse responses for $H(z)$, $H_r(z)$, and $H_t(z)$. Next compute the error signals:

$$e_t[n] = h[n] - h_t[n]$$

and

$$e_r[n] = h[n] - h_r[n]$$

using the **f subtract** function. Using **g view2**, display the plots of these error signals and observe the distortion. Comment on any differences that you observe. Quantization error is often modeled as a random noise source. Do your observations support the notion that random noise sources are reasonable models for quantization error?   □

EXERCISE 5.3.3.   **Coefficient Sensitivity of Cascade Structures**

Quantization of direct form filter coefficients can significantly perturb the poles and zeros. This exercise considers the eighth-order bandpass filter stored in the file **h0df**. Use **g polezero** to display and sketch the magnitude response and pole/zero plot for this filter.

The filter can be represented in cascade form. The cascade form coefficients for the four second-order sections that, when cascaded, are equivalent to **h0df** are stored in the DSP files **h0c1**, **h0c2**, **h0c3**, and **h0c4**. The point of this exercise is to compare the effect of quantizing the filter coefficients in direct form and cascade form.

(a) Use **f convert** to split **h0df** into numerator and denominator files. Separately quantize these coefficient files to 2048 levels (11 bits) using **f quantize**. X_{min} and X_{max} should be -0.013126 and 0.0180014 for the numerator and -4.594558 and 5.739664 for the denominator, respectively. Reconstruct the quantized filter using **f revert** and display and sketch the magnitude response using **g polezero**. You should observe some degradation due to the quantization.

(b) Now apply the same procedure to the cascade form filters **h0c1**, **h0c2**, **h0c3**, and **h0c4**. In particular, examine each second-order section. You will observe that -1.22528 and 0.733166 are the minimum and maximum values for the numerators and -0.75465 and 1 are the minimum and maximum values for the denominators.[8] Use these values as quantization limits for each of the sections.

---

[8]Remember that, in the case of the denominator, the leading coefficient is always fixed at one and does not appear in the DSP file. Thus the maximum value for the denominator can never be smaller than one.

Begin the quantization procedure by converting each second-order section into a numerator and denominator file. Quantize each to 2048 levels using the minimum and maximum values stipulated above. Then use **f revert** to recombine the quantized numerators and denominators. The last step is to convert the sections into direct form by repeatedly using **f cas** to cascade the sections into one direct form filter so that it can be displayed. You may wish to put these commands in a macro. This may make it easier to keep track of your intermediate files. Display and sketch the pole/zero plot and magnitude response of this filter. It corresponds to the quantized cascaded realization.

(c) Provide an explanation based on the clustering of poles and zeros for why the cascade form realization is less sensitive to coefficient quantization than the direct form. □

EXERCISE 5.3.4.   Sensitivity of Poles Near the Unit Circle

We have seen that the locations of closely clustered poles are sensitive to perturbations in the direct form coefficients. When poles are near the unit circle, the filter frequency response is often sensitive to their locations.

(a) Using **g polezero**, display and sketch the pole/zero plot and magnitude response for $H_2(z)$ found in the DSP file **h2df**. Use the *change pole/zero* option to perturb the pole pair closest to the unit circle. Specifically, move the pole pair one increment away and then back again in each direction. Display the magnitude response interactively after making each change by pressing the "d" key. Sketch the magnitude response resulting from each purturbation.

(b) Now perturb the other two pole pairs in the same way indicated in part (a). Display and sketch the changes in the magnitude response.

When coefficient quantization causes poles close to the unit circle to move, the impact on the magnitude response can be severe. □

EXERCISE 5.3.5.   Parallel Form Coefficient Sensitivity

Filters can be expressed in parallel form to yield implementations that are more resilient to coefficient quantization than direct form realizations. In this exercise, we consider the eighth-order bandpass filter stored in the filter file **h0df**. Display and sketch the magnitude and pole/zero plots of this filter using **g polezero**.

This filter can be represented in parallel form using the filter sections that are provided in the files **h0p1**, **h0p2**, **h0p3**, **h0p4**, and **h0p5**. All of these sections are second order with the exception of the last, which is a constant term.

In this exercise, we compare the effects of coefficient quantization in the direct form and parallel form realizations. We begin, in part (a), by quantizing the direct form filter **h0df**. If you have worked Exercise 5.3.3a, then you have already quantized **h0df** and examined it. In this case, redraw the magnitude response of the quantized filter and proceed to part (b).

(a) Use **f convert** to split **h0df** into numerator and denominator files. Separately quantize these coefficient files to 2048 levels (11 bits) using **f quantize**. X_{min} and X_{max} should be -0.016758 and 0.0189746 for the numerator and -4.61069 and 5.7782

for the denominator, respectively. Reconstruct the quantized filter using **f revert** and display and sketch the magnitude response using **g polezero**. You should observe some degradation due to the quantization.

(b) Now apply the same quantization procedure to the parallel form filters. Examine the filter sections **h0p1**, **h0p2**, **h0p3**, **h0p4**, and **h0p5** and observe that -0.33484 and 0.297716 represent the minimum and maximum values for the set of numerators and -0.754629 and 1.00 represent minimum and maximum[9] values for the denominators. Use these values as quantization limits for each of the sections. The section **h0p5** is actually a constant and not a second-order section. Therefore, do not quantize the denominator of the section, only the numerator.

Use **f revert** to recombine the numerators and denominators. The last step is to convert the sections into direct form for display using the **f par** function. This function can be used repeatedly to merge the sections into one direct form filter. It might be more convenient to put these commands in a macro. This will make it easier to keep track of your intermediate files.

Display and sketch the pole/zero plot and magnitude response of this filter. This corresponds to the quantized parallel form realization.

(c) Explain why the parallel form realization is less sensitive to coefficient quantization than the direct form based on the clustering of poles and zeros.   □

EXERCISE 5.3.6.   **Coefficient Sensitivity in Allpass Filters**

As discussed in Chapter 4, allpass filters can be implemented with relatively few multiplications. For example, an allpass filter of order N can be implemented with only N multiplications using the difference equation

$$y[n] = a_1(x[n - (N - 1)] - y[n - 1]) + a_2(x[n - (N - 2)] - y[n - 2])$$
$$+ \cdots + a_N(x[n] - y[n - N]) + x[n - N], \tag{5.35}$$

where $x[n]$ is the input sequence, $y[n]$ is the output sequence, and $a_1, a_2, \ldots, a_N$ are the allpass filter coefficients. A difference equation of this form requires approximately half the multiplications of a direct form implementation. There are other implementations for allpass filters that are equally efficient in terms of a minimum number of multiplications. All of these have the property that they maintain their allpass characteristic even when the filter coefficients are severely quantized. In this exercise, we will examine the effects, on the frequency response, of quantizing allpass filters and filters based on the sum of allpass filters.

Using a text editor, create the following macro called **quan.bat** that will split an IIR filter into its numerator and denominator polynomials, quantize each separately, and merge them back.

---

[9]Remember that 1 is always the first coefficient in the denominator polynomial but is not explicitly shown in the DSP file format. In this case, it is the maximum coefficient value.

> **f convert   %1   temp   temp1**
> **f quantize   temp   temp2   −%3   %3   %4**
> **f quantize   temp1   temp3   −%3   %3   %4**
> **f revert   temp2   temp3   %2**

For convenience, we will assume the quantization range to always be symmetric about zero, i.e., $-X_{max}$ to $+X_{max}$. The macro allows you to enter X_{max} and the number of quantization levels on the command line.

(a) Consider the fourth-order allpass filter stored in the file **h3dfall1**. Type the filter to the screen and note the dynamic range of coefficient values. Quantize this filter to 64 levels (6 bits) in the range -1 to $+1$ by typing

> **quan   h3dfall1   out1   1   64**

Observe that additional distortion is introduced because the maximum value of **h3dfall1** is 1.1736 and not 1.

Use **g polezero** to examine **h3dfall1** and **out1**. Sketch the pole/zero and magnitude response plots for each. Note that in spite of the severe quantization, the quantized filter remains allpass.

(b) This coefficient quantization property of allpass filters has implications for frequency selective filters that are composed of sums of allpass filters. Consider the bandpass filter $H_3(z)$ stored in the file **h3df**. Display and sketch it using **g dtft**. This filter can be written as the sum of two allpass filters, $\mathcal{A}_0(z)$ and $\mathcal{A}_1(z)$, as discussed in Chapter 4. The allpass filters $\mathcal{A}_0(z)$ and $\mathcal{A}_1(z)$ are stored in the files **h3dfall1** and **h3dfall2**, respectively. You will now compare the frequency response of the direct form and allpass form implementations of $H_3(z)$ when the coefficients are quantized.

First, scale $H_3(z)$ by a factor of 60 using the **f gain** function. This will make the dynamic range of the numerator and denominator about the same and approximately equal to -4.5 to 4.5. Use **quan.bat** to quantize the scaled version of $H_3(z)$ to 10 bits (1024 levels) with $X_{max} = 4.5$. Display and sketch the magnitude response of the scaled original and the quantized version using **g dtft** and observe the degradation in the passband.

Now quantize $\mathcal{A}_0(z)$ and $\mathcal{A}_1(z)$ to 10 bits. Since the dynamic range of $\mathcal{A}_0(z)$ and $\mathcal{A}_1(z)$ is only ± 2, use $X_{max} = 2$ as the range parameter. Add the quantized allpass filters together using the **f par** function. Display and sketch the resulting magnitude response using **g dtft**. Notice the improvement in performance.

(c) Systematically lower the number of bits used to represent $H_3(z)$ in the allpass implementation (using $X_{max} = 2$) until the magnitude response characteristics deteriorate to a point comparable to that of the direct form implementation at 10 bits. How many bits are required with the allpass implementation to provide a comparable level of performance?   □

EXERCISE 5.3.7.   Coefficient Quantization in FIR Filters

Coefficient quantization in FIR filters is much easier to analyze than for IIR filters. Quantization of the direct form coefficients $h[n]$ for a given FIR filter results in

$$\hat{h}[n] = h[n] + \Delta h[n],$$

where $\Delta h[n]$ is the coefficient quantization error. This leads to a frequency response

$$\hat{H}(e^{j\omega}) = H(e^{j\omega}) + \Delta H(e^{j\omega}).$$

Design a 31-point FIR lowpass filter using the *Hamming window* option in **f fdesign** with cutoff frequency $0.65\pi = 2.042$. Quantize the coefficients to 1024 levels using the **f quantize** function. Examine the coefficients, preferably using a text editor, and verify that the minimum and maximum coefficient values are approximately -0.12 and 0.65, respectively. Use these values for X_{min} and X_{max} in **f quantize**.

(a) Compute the difference signal $\Delta h[n] = \hat{h}[n] - h[n]$ using **f subtract**. Display and sketch the difference signal $\Delta h[n]$ using **g view**. Observe the random nature of this signal. Now display and sketch $|\hat{H}(e^{j\omega})|$ using **g dtft**.

(b) Compare the degradation in the magnitude response due to quantization of the co-efficients of the FIR filter for 1024-level quantization with that of the IIR lowpass filter $H_2(z)$ in Exercise 5.3.1 with 2048-level quantization. Provide a sketch of both. Which appears to do a better job of preserving the magnitude response when the coefficients are quantized (in spite of the higher-precision IIR representation)?   □

EXERCISE 5.3.8.   **Coefficient Quantization in FIR Filters**

A quantized FIR filter, $\hat{H}(z)$, can be expressed in terms of an unquantized component and an error component, $\Delta H(z)$:

$$\hat{H}(z) = H(z) + \Delta H(z).$$

Taking the DTFT of the error term $e[n] = h[n] - \hat{h}[n]$ results in

$$\Delta H(e^{j\omega}) = \sum_{n=0}^{M} e[n]e^{j\omega n},$$

where $\Delta H(e^{j\omega})$ is the frequency error. Since $e[n]$ is bounded by $\pm\delta/2$ (where δ is the quantization step size), the FIR filter quantization error is bounded, i.e.,

$$|\Delta H(e^{j\omega})| \le (M+1)\delta/2. \tag{5.36}$$

In this exercise, we examine the quantization of FIR filters, observe the effects of the distortion, and compare the distortion with the bound in (5.36).

Design a 64-point FIR lowpass filter, $h[n]$, with cutoff frequency 1.57 rad using a Kaiser window with parameter $\alpha = 2$. Use the functions **f ideallp** to design the ideal prototype filter, **f kaiser** to design the Kaiser window, and **f multiply** to multiply the window and prototype together. Display and sketch the magnitude and log magnitude responses of this filter using **g dfilspec**.

Assume that the system specifications require that the magnitude response of the quantized filter satisfy the passband and stopband conditions: $\delta_s < 0.06$ and $\delta_p < 0.063$. Using the **f quantize** function with quantization range -1 to $+1$, determine the minimum number of bits necessary to represent $h[n]$.   □

5.4   FINITE PRECISION NUMBER REPRESENTATION

Hardware always imposes limits on the numerical word lengths that can be used to represent signals and coefficients. Consequently, the effects of finite precision arithmetic are important considerations for filters that are to be implemented in hardware. In practice, input data samples are quantized to a predefined number of levels. If a two's complement representation is used, each level is then represented by a binary word having the form

$$x = (\text{mantissa}) \ 2^{\text{(characteristic)}},$$

where

$$\text{mantissa} = \underbrace{-b_0}_{\text{sign bit}} + \sum_{\ell=1}^{B} b_\ell 2^{-\ell}, \quad b_0, b_1, \ldots, b_B = 0, 1,$$

and the *characteristic* or *exponent* is also a binary number. When the characteristic is fixed, the representation is called *fixed point*. The mantissa then assigns to each level in the quantizer a binary number. Scaling, which effectively adjusts the dynamic range of the quantizer to match the signal, is an important operation in fixed-point systems.

A very large dynamic range can be achieved when a variable characteristic is used. This is called a *floating-point* representation. Both floating-point and fixed-point representations have numerical errors associated with them; however, these errors can be kept small in many cases if the proper precautions are taken. Floating-point representations, in general, are easier to work with because of their larger dynamic range and the lack of a need to scale. Fixed-point arithmetic, on the other hand, has a far more limited dynamic range, but has the advantage that it is structurally very simple and is thus much less demanding in terms of hardware. The thrust of the treatment in this text is on numerical errors in fixed-point systems. Understanding these issues is important because fixed-point systems can be numerically sensitive.

The binary number representation used in a system must be able to represent both positive and negative numbers and must easily accommodate the basic arithmetic operations. Two's complement is the most popular format, and it is the only one that we will consider. To accommodate multiplications conveniently, samples and filter coefficients are typically represented as binary fractions. In this way, the product of two (B+1)-bit numbers is always a binary fraction. This convenient convention prevents multiplication operations from producing products that exceed the range of the number representation. Truncation or rounding after multiplications is used to maintain a constant word length. This is a major source of quantization error in numerical processing, but the magnitude of the error is limited to the magnitude of the least significant bit.

When two binary fractions are added, the sum can exceed one, resulting in a number that is no longer a binary fraction. This situation is called *overflow* and is often handled by scaling. If the system is not properly scaled, overflow errors can be quite large, which constitutes a major source of errors. These issues and others related to number representations are treated in the exercises that follow.

EXERCISE 5.4.1. Two's Complement Representations

A positive number, α, may be converted to a negative number, $-\alpha$, in two's complement notation by subtracting its magnitude from 2. (Note that 2 in binary equals 10.0000.) The same effect can be achieved by complementing all of the bits in the magnitude, adding one to the least significant bit and adding a one for the sign bit. We start by assuming an 8-bit fractional binary representation for the numbers.

(a) Express 7/8 and $-7/8$ in two's complement form.

(b) If $x = 0.00111000$ and $y = 1.01010110$ are two's complement numbers, what are their values?

(c) Two's complement arithmetic has a special property with respect to overflow, which can be illustrated as follows. Consider the sum of three numbers: $3/4+3/4+(-7/8) = 5/8$. First we add 3/4 and 3/4, which, in binary, gives $0.110+0.110 = 1.100$. Since the sum is greater than one, it is in a state of overflow. What is the value of 1.100 in the two's complement representation? Now add the two's complement equivalent of $-7/8$ to the sum. What is the numerical value of the result?

   This is a special property of two's complement arithmetic. As long as the output value is within the range of the representation, two's complement addition will always give the correct result, even if overflow occurs in the intermediate calculations. □

EXERCISE 5.4.2. Wraparound Overflow

This exercise considers the property of two's complement arithmetic that was observed in the previous exercise. Two's complement arithmetic has the special property, with respect to addition, that a sum whose magnitude exceeds the range of the number representation is replaced by its value modulo the range of the quantizer. As long as the final output sum is within the range of the representation, the correct answer will always be produced.

   The *wraparound* arithmetic overflow property of the two's complement representation is implemented in the function **f qcyclic**. In this exercise, we will examine this cyclic or wraparound effect visually. The operator, $Q_c\{\cdot\}$, will be used to denote *wraparound* quantization and $Q\{\cdot\}$ will denote conventional quantization (or quantization with saturation). Use **f siggen** to design four signals:

   (i) an impulse, $\delta[n]$;

   (ii) a 10-point ramp, $r[n]$, with slope=0.08 and starting point $n = 0$;

   (iii) a 10-point block, $b[n]$, with amplitude=0.5 and starting point $n = 0$; and

   (iv) a 10-point sinusoid, $s[n]$, with amplitude=0.5, alpha = 0.1, phi = π, and starting point $n = 0$.

Assume a numerical sample representation consisting of 31 levels that are uniformly spaced between -1 and $+1$. These conditions are to determine the parameter settings that you will use for **f quantize** and **f qcyclic** throughout this exercise.

(a) Begin by applying **f quantize**, as described above, to each of the four signals to form

$$\hat{\delta}[n] = Q\{\delta[n]\},$$

$$\hat{r}[n] = Q\{r[n]\},$$

$$\hat{b}[n] = Q\{b[n]\},$$

$$\hat{s}[n] = Q\{s[n]\}.$$

Use **g view2** to display the quantized and unquantized signals on the screen and observe the effects of the quantization.

To simulate two's complement arithmetic, numbers outside the interval $(-1, +1)$ must exhibit wraparound. The **f qcyclic** function implements the $Q_c\{\cdot\}$ operation. To investigate this function, use **f gain** to form the impulse

$$\delta_1[n] = 1.25\hat{\delta}[n].$$

Next use **f qcyclic** to perform 31-level quantization on $\delta_1[n]$, i.e.,

$$\hat{\delta}_1[n] = Q\{\delta_1[n]\}$$

with range -1 to $+1$ and also on $\hat{\delta}[n]$. Use **g view2** to display $\hat{\delta}_1[n]$ and $\hat{\delta}[n]$. Notice that because $\hat{\delta}[n]$ lies within the interval $(-1, +1)$, it is undisturbed by the $Q_c\{\cdot\}$ operation. However, $\hat{\delta}_1[n]$ is affected.

Using **f add** and **f qcyclic**, compute $\hat{v}[n] = Q_c\{\hat{r}[n] + \hat{b}[n]\}$. Display and sketch $\hat{v}[n]$. Now compute $\hat{y}[n] = Q_c\{\hat{v}[n] + \hat{s}[n]\}$ and sketch the result.

(b) Compute the same signals as in part (a) but this time omitting the wraparound quantization after the additions by simply computing $v[n] = \hat{r}[n] + \hat{b}[n]$ and $y[n] = v[n] + \hat{s}[n]$ using the **f add** function. Use **g view2** to display $v[n]$ and $\hat{v}[n]$ and sketch the signals. Next use **g view2** to display $y[n]$ and $\hat{y}[n]$ and sketch them as well. Explain why $v[n]$ and $\hat{v}[n]$ are different, yet $y[n]$ and $\hat{y}[n]$ are the same. Save these signals files for use in the next exercise. It continues with a discussion of saturation arithmetic, which is an alternative method for handling overflow. □

EXERCISE 5.4.3. Saturation Versus Wraparound Overflow

The addition of two numbers in a bounded finite number representation can result in overflow. The previous exercise looked at one type of overflow, wraparound overflow. You should complete that exercise before beginning the current one. Here we examine another method for handling overflow called *saturation*. In this approach, samples that exceed the upper quantization range limit are fixed to the maximum representable value and samples that fall below the lower range limit are assigned the minimum range value. Thus, values that overflow during the evaluation of an arithmetic expression will result in an output of either X_{min} or X_{max}. Saturation can have advantages over wraparound overflow.

(a) To simulate saturation arithmetic, we will use the sequences $\hat{r}[n]$, $\hat{b}[n]$, and $\hat{s}[n]$ defined in the previous exercise and quantize these signals with a saturating quantizer $Q\{\cdot\}$ as described above. Compute $\tilde{v}[n] = Q\{\hat{r}[n] + \hat{b}[n]\}$ and $\tilde{y}[n] = Q\{\tilde{v}[n] + \hat{s}[n]\}$ using **f add** and **f quantize** with 31 levels and with a minimum and maximum quantization range of -1 and $+1$, respectively.

Using **g view2**, display and sketch $\tilde{v}[n]$ and $\tilde{y}[n]$, which you just computed, and $v[n]$, $y[n]$, $\hat{v}[n]$, and $\hat{y}[n]$ from the previous exercise. Why are errors present when saturation arithmetic is used?

Now repeat the evaluation of $\tilde{y}[n]$ with a modification of the order of operations: $\tilde{v}[n] = \hat{r}[n] + \hat{s}[n]$ and $\hat{y}[n] = \hat{v}[n] + \hat{v}[n]$. How has the change in order improved the result?

(b) Now consider comparing saturation and wraparound arithmetic. Compute the sum of $1.5\hat{r}[n]$ and $\hat{b}[n]$ first using wraparound arithmetic and then using saturation arithmetic, i.e., compute

$$\hat{z}[n] = Q_c\{1.5\hat{r}[n] + \hat{b}[n]\}$$

and

$$\tilde{z}[n] = Q\{1.5\hat{r}[n] + \hat{b}[n]\}.$$

Display and sketch the results. In this case, some samples of the true result exceed the quantization range. How do wraparound and saturation compare when the output overflows?   □

5.5 ROUNDOFF EFFECTS IN FILTER IMPLEMENTATION

When two (B+1)-bit numbers are multiplied, the product is a 2B+1 bit number. In order to maintain constant word lengths for samples as they pass through the system, rounding or truncation is applied to restore intermediate products to a (B+1)-bit representation, which introduces error into the system that can adversely affect the filter's performance. We would often like to determine the minimum word length necessary to keep the quantization error within reasonable limits. We will begin the discussion with a linear model that can be used to represent roundoff and truncation noises in a system and later examine it in the context of the classical implementation structures.

Figure 5.8a illustrates multiplication of an input sample by a constant a using infinite precision arithmetic. Figure 5.8b illustrates the same multiplication when roundoff is applied after the multiplication. Both the input signal and product signal are quantized to maintain constant wordlength. Because of quantization, the true system is nonlinear and is, therefore, very difficult to analyze. However, quantization error can be modeled reasonably well as a wide-sense stationary white noise source. This leads to the linear approximation or additive noise model shown in Figure 5.8c. Modeling an arbitrary structure in this way is done by replacing all of the quantization operators with additive noise sources throughout the flow graph. If the noise sources are statistically independent and spectrally white, they can be described by their means and variances. For the case of rounding to a B+1 bit wordlength, the mean is zero and the probability density function or pdf can be assumed to be reasonably uniformly distributed between $-2^{-B}/2$ and $+2^{-B}/2$. The variance of such a random variable is $2^{-2B}/12$. The output, due strictly to these noise sources, can now be examined.

(a)

(b)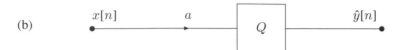

(c)

Figure 5.8. (a) Flow graph for an ideal multiplier. (b) Flow graph for a multiplier with a quantized product. (c) Flow graph of a linear model for a multiplier with a quantized product.

Figure 5.9 shows the flow graph for a direct form I implementation of a second-order system. Inspection of the flow graph shows that the five noise sources can be combined into one as shown in Figure 5.10, where

$$e_T[n] = e_1[n] + e_2[n] + e_3[n] + e_4[n] + e_5[n].$$

Since these sequences are assumed to be independent, identically distributed white noise sources and uncorrelated with the input, the variance of $e_T[n]$ is the sum of the variances of the individual noise sources, i.e.

$$\sigma_T^2 = \sigma_1^2 + \sigma_2^2 + \sigma_3^2 + \sigma_4^2 + \sigma_5^2 = 5\frac{2^{-2B}}{12},$$

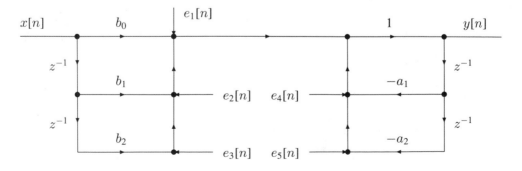

Figure 5.9. Direct form I implementation of a second-order section with quantization modeled by the inclusion of noise sources.

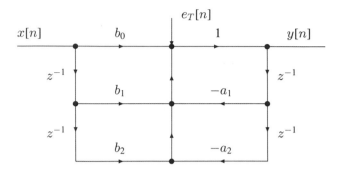

Figure 5.10. Direct form I implementation of a second-order section with quantization modeled by one combined noise source.

where $\sigma_1^2 = \sigma_2^2 = \ldots = \sigma_5^2 = 2^{-2B}/12$ are the noise source variances. With a little thought, however, we see that if the summations in the center of the structure are performed using $(2B + 1)$-bit arithmetic, only one quantization, rather than five, needs to be performed. This will reduce the variance of $e_T[n]$ by a factor of five. (See Exercise 5.5.5.)

If the input to the ideal system is $x[n]$ and the output is $y[n]$, then the output of a real system with roundoff or truncation noise is $y[n] + \Delta y[n]$, where $\Delta y[n]$ is the output component due to the quantization noise. Since our analysis model is linear, we can express $\Delta y[n]$ in terms of $e_T[n]$. In other words, we can express the output error $\delta y[n]$ as a function of the input noise source $e_T[n]$. This relationship leads to the system function for the quantization error, $F(z)$, where

$$F(z) = \frac{\Delta Y(z)}{E_T(z)}.$$

The impulse response of $F(z)$ is $f[n]$, which can be used to measure the sensitivity of the structure.

A very useful measure of the sensitivity of a structure to quantization error is the output noise variance $\sigma_{\Delta y[n]}^2$. For a direct form I implementation,

$$\sigma_{\Delta y[n]}^2 = \sigma_T^2 \sum_{n=-\infty}^{\infty} |f[n]|^2 \tag{5.37}$$

or equivalently

$$\sigma_{\Delta y[n]}^2 = \sigma_T^2 \frac{1}{2\pi} \int_{-\pi}^{\pi} |F(e^{j\omega})|^2 d\omega$$

due to Parseval's relation. The smaller this variance, the less sensitive the system is to rounding and truncation error.

A direct form II implementation can be analyzed in a similar manner. In this case, however, the structure is such that the noise sources are combinable into two separate composite noise sources instead of just one. The first noise component occurs at the input node of the direct form II flow graph, the second is at the output node. This leads to the output noise variance expression,

$$\sigma^2_{\Delta y[n]} = N\sigma^2_0 \sum_{n=-\infty}^{\infty} |h[n]|^2 + (M+1)\sigma^2_0, \qquad (5.38)$$

where $\sigma^2_0 = 2^{-2B}/12$, $h[n]$ is the impulse response of the entire filter N is the number of poles, and M is the number of zeros. These expressions for the direct form I structure (5.37) and direct form II structure (5.38) can be used to estimate the necessary data wordlengths given a specified output noise variance.

Several points become clear upon examining this quantization analysis model. First, the output noise variance is typically inversely related to the distance of the poles from the unit circle. A system with poles close to the unit circle is more sensitive to quantization noise. Second, the output noise variance is inversely related to wordlength. Thus, numerical sensitivity can be reduced by increasing the processing wordlength. Finally, this analysis method can be used to estimate useful statistics such as the error mean and noise variance. The accuracy of the model is not very good when the input is a narrow band signal, such as a sinusoid, but is reasonable when the input is a wideband signal with good amplitude variation.

Two very useful functions are provided in the software for examining the effects of finite precision processing distortion: **f direct1** and **f direct2**. They implement the difference equations corresponding to the direct form I (5.2) and direct form II (5.5) and (5.6), respectively. Type **f direct1** and press the "enter" key. It will display the difference equation that it implements. Press "control c" to exit at this point. Now do the same for **f direct2**.

The input parameters to these functions can also be input on the command line by typing, for example,

f direct1   xn   H(z)   out1   out2   −1   +1   1024   0   200

The first parameter **xn** is the input sequence; $H(z)$ is the filter; **out1** is the approximate ideal output, computed with double precision floating-point arithmetic; and **out2** is the output, computed in the fixed-point processing environment. The next two parameters are the minimum and maximum quantizer range for the implementation. In this case it is −1, +1. The following parameter is the number of quantization levels, 1024 in this case. The next parameter is either a 0 or 1, 0 for saturation arithmetic and 1 for wraparound arithmetic. The last parameter is the number of samples to be computed in the output. In the example, 200 output samples will be computed. A special verbose mode of operation occurs when 0 is specified for this parameter. When this is done, the function will display a detailed chart of the multiplication operations, quantized products, and sum of intermediate terms that are computed in the computation process for each value of n. The function will first display the chart for $n = 0$. Pressing the "c" key will display the chart for $n = 1$ and so on.

We will make extensive use of these functions in our study of overflow, scaling, and limit cycles.

EXERCISE 5.5.1. **Estimating the Data Wordlength**

A common objective in roundoff error analysis is to determine the minimum wordlength (in bits) necessary to keep the noise power relatively small within a given structure.

(a) Consider the implementation of a second-order filter, $H(z)$, in a direct form I implementation with a signal quantization range from -1 to $+1$, where

$$H(z) = \frac{1}{1 - (4/9)z^{-2}}.$$

Write the expression for the output noise variance predicted by (5.37) in terms of B. Notice that $B + 1$ is the number of bits in the fixed-point representation. Based on this equation, how many bits should be used to keep the noise power below 8×10^{-6}?

(b) Generate a 200-point random input sequence, $x[n]$, that begins at $n = 0$ using **f rgen** with an initial seed value of 6. Next, use the *create file* option in **f siggen** to create a file for $H(z)$. Evaluate 200 output samples of this filter with finite precision wraparound arithmetic using a direct form I implementation. This may be done by typing

$$\textbf{f direct1} \quad \textbf{xn} \quad \textbf{H(z)} \quad \textbf{out1} \quad \textbf{out2} \quad -1 \quad +1 \quad (2^{\textbf{B+1}} - 1) \quad 1 \quad 200$$

where $B + 1$ is the wordlength of the computation that was estimated in the previous part. To obtain a quantizer that represents zero using the **f direct1** function, we specify an odd number of levels. This is why $2^{B+1} - 1$ and not 2^{B+1} is used for the number of levels. The file **out2** will contain 200 samples of the output evaluated with finite precision wraparound arithmetic (i.e., $\hat{y}[n]$) while the file **out1** will contain the output computed using floating-point arithmetic, which can be used as the ideal output, $y[n]$.

The output error signal, $\Delta y[n]$, may be defined as $\Delta y[n] = \hat{y}[n] - y[n]$ and may be computed by using the **f subtract** function to subtract **out1** from **out2**. Explicitly evaluate the output noise variance

$$\sigma^2_{\Delta y[n]} = \frac{1}{200}\sum_{n=0}^{199}|\Delta y[n]|^2 - \left(\frac{1}{200}\sum_{n=0}^{199}\Delta y[n]\right)^2,$$

much of which can be done using the **f summer** function. Is the noise variance below the 8×10^{-6} limit?

Although the accuracy of the noise variance expressions may vary from filter to filter, these expressions can be very useful in determining data wordlengths for fixed-point systems. □

EXERCISE 5.5.2. **Effects of Rounding in Direct Form II Filters**

Consider the analysis for the direct form II implementation of the filter

$$H(z) = \frac{0.3333 + 0.6667z^{-1}}{1 - 0.5z^{-1} - 0.1z^{-2}}.$$

(a) Sketch the direct form II flow graph with noise sources inserted in place of the quantizers. Next, combine the noise in the feedback (fb) paths to form $e_{fb}[n]$ and the sources in the feedforward (ff) paths to form $e_{ff}[n]$. This results in three input components to the system, $x[n]$, $e_{fb}[n]$, and $e_{ff}[n]$, each with a corresponding output $y[n]$, $y_{fb}[n]$, and $y_{ff}[n]$, respectively. Determine $F_{fb}(z) = Y_{fb}(z)/E_{fb}(z)$ and $F_{ff}(z) = Y_{ff}(z)/E_{ff}(z)$.

The output error signal, $\Delta y[n]$, is defined as $\Delta y[n] = \hat{y}[n] - y[n]$, the difference between the fixed-point and ideal output signals. Show that (5.38) is a valid expression for the output noise variance $\sigma^2_{\Delta y[n]}$. Assuming a $B + 1 = 9$ bit wordlength, determine $\sigma^2_{\Delta y[n]}$ predicted by the additive noise model.

(b) Now generate a 200-point random input sequence file, **xn**, that begins at $n = 0$ using **f rgen** with initial seed value of 0. Use the *create file* option in **f siggen** to create a file for $H(z)$. Evaluate 200 output samples using finite precision arithmetic for a direct form II implementation with saturation arithmetic. This may be done by typing

$$\textbf{f direct2} \quad \textbf{xn} \quad \textbf{H(z)} \quad \textbf{out1} \quad \textbf{out2} \quad -1 \quad +1 \quad 512 \quad 0 \quad 200$$

where **out2** will contain 200 samples of the output evaluated with finite precision saturation arithmetic and **out1** will contain the output evaluated using floating-point arithmetic.

Next compute the output error signal, $\Delta y[n]$, using the **f subtract** function to take the difference between **out2** and **out1**. Explicitly evaluate the output noise variance

$$\sigma^2_{\Delta y[n]} = \frac{1}{200}\sum_{n=0}^{199}|\Delta y[n]|^2 - \left(\frac{1}{200}\sum_{n=0}^{199}\Delta y[n]\right)^2,$$

much of which can be done using the **f summer** function. In terms of a percentage difference, how close is the agreement between the predicted noise variance and the variance that you actually computed?   □

EXERCISE 5.5.3.   **Performance of Cascade Form Filters**

The error analysis discussed previously can be a useful tool but it does not have the accuracy to determine which, among a set of candidate implementation structures, will produce the smallest error for a given input. Here we examine the cascade structure and compare it to the direct form I structure in terms of its performance in a fixed-point environment.

Begin by creating a 200-point random input sequence, $x[n]$, that begins at $n = 0$ using **f rgen** with an initial seed value of 0. This input will be filtered by the lowpass filter given in Table 5.1 that is contained in the DSP file **f010** provided in the software. Evaluate 200 output samples assuming a signal quantization range of -1 to 1, 512 quantization levels, and saturation arithmetic. This may be done by typing

$$\textbf{f direct1} \quad \textbf{xn} \quad \textbf{H(z)} \quad \textbf{out1} \quad \textbf{out2} \quad -1 \quad +1 \quad 512 \quad 0 \quad 200$$

where **out2** will contain 200 samples of the output evaluated with finite precision saturation arithmetic and **out1** will contain the output computed using floating-point arithmetic.

Now consider evaluating the same filter implemented as a cascade of two second-order direct form II sections. These cascade form sections are given in Table 5.1 and are also in the files **f010c1** and **f010c2**. First filter $x[n]$ with **f010c1** to obtain 200 output samples. Then use the quantized output of this section as input to the filter **f010c2** and generate 200 output samples. Use the **f direct2** function for filtering and the same quantization specifications as before.

Use the unquantized output from the direct form I implementation as a representation of the ideal output. Compute an error signal $\Delta y_c[n]$ for the cascade implementation and an error signal $\Delta y_d[n]$ for the direct form I implementation by subtracting each of the quantized signals from the ideal using **f subtract**. Display $\Delta y_c[n]$ and $\Delta y_d[n]$ using **g view2** and again using **sview2**. Which implementation appears to be the better one to use for this filter?   □

EXERCISE 5.5.4. **Performance of Parallel Form Filters**

Repeat the previous exercise for a parallel form structure using **f direct2** to implement each second-order section. Here it will be necessary to use $x[n]$ as input to all three sections and the **f add** function to sum all the section outputs together. Files containing the parallel form section filters are contained in **f010p1**, **f010p2**, and **f010p3**.   □

EXERCISE 5.5.5. **Improving Performance**

One solution to improving the processing accuracy in direct form, cascade, and parallel implementations is to use double-precision registers to accumulate the sums of intermediate products. A double-precision register has twice the length of the numerical word, and thus, when used to store the product of two numbers, does not introduce error. After products are accumulated in this register, the double-precision result may then be rounded to a single-precision wordlength. In terms of our error analysis, this procedure has the effect of removing the quantizers that follow multiplication operations and replacing them with a single quantizer after the summation node.

Assume that a wordlength is $B + 1$ bits and that the accumulator register is $2B + 1$ bits. Given a general tenth-order digital filter, with 10 poles and 10 zeros, analytically determine the output noise variance as a function of B for a direct form I implementation. How does this implementation compare to one that uses a single precision accumulator register?   □

EXERCISE 5.5.6. **Performance of Allpass Filters**

In this exercise, we examine the bandpass filter $H_3(z)$ stored in the file **h3df**, first implemented in direct form and then implemented as a sum of two allpass filters.

(a) Use the **f direct1** function with the file **impulse** as input and compute 100 samples of the impulse response of $H_3(z)$ with the quantization range, -3 to $+3$, the number of quantization levels, 512, and wraparound arithmetic. This may be done by typing

    f   direct1   impulse   h3df   out1   out2   −3   3   512   1   200

The file **out1** contains the output evaluated with floating-point arithmetic while **out2** is the result of using finite precision arithmetic. Display and sketch both outputs using **g view2**. The distortion you observe is due to quantization effects in the evaluation of the difference equation. Use **g dtft** to display the frequency responses for **out1** and **out2** and observe that the bandpass characteristics have been severely degraded due to the finite precision arithmetic.

(b) The filter $H_3(z)$ can also be implemented as a sum of two allpass filters, in particular the sum of those stored in files **h3dfall1** and **h3dfall2**. Explicitly evaluate the impulse responses for both allpass filters under fixed-point and floating-point arithmetic and under the same quantization conditions as before by typing

> **f direct1   impulse   h3dfall1   a1   qa1   −3   3   512 1 200**

> **f direct1   impulse   h3dfall2   a2   qa2   −3   3   512 1 200**

Use **f add** to sum the outputs **a1** and **a2** to obtain the floating-point output **hn**. Now sum the outputs **qa1** and **qa2** to obtain the fixed point output **qhn**. Display and sketch **hn** and **qhn** using **g view2**. Notice that the time domain characteristics of the impulse response are better preserved when the implementation is done as a sum of allpass sections. Now examine the DTFT magnitude of **qhn** using **g dtft**. Notice that the frequency domain behavior is better also.   □

5.6   OVERFLOW EFFECTS AND SCALING

A potentially dominant source of distortion in a fixed-point processing environment is overflow. It occurs when the sum of two or more numbers falls outside of the range limits of the number representation. Overflow can be prevented by scaling the signal levels at certain summation nodes in the implementation. A necessary and sufficient condition to guarantee that each node value is less than one is that

$$X_{max} < \frac{1}{\left(\max_{k} \sum_{m=-\infty}^{\infty} |f_k[m]| \right)}, \tag{5.39}$$

where X_{max} is the maximum magnitude of the input and $f_k[n]$ is the impulse response of the system at node k, i.e., the response obtained at node k, when the input to the system is $\delta[n]$. Remember that nodes in a flow graph are those points at which intermediate terms are accumulated. For example, in the case of a direct form II implementation of $H(z) = B(z)/A(z)$, there is one node in the center of the flow graph with difference equation

$$w[n] = x[n] - a_1 w[n-1] - \cdots - a_N w[n-N].$$

Thus $f_k[n]$ (where $k = 1$) is just the impulse response of the feedback section of the flow graph, i.e.,

$$1/A(z) = 1/(1 + a_1 z^{-1} + \cdots + a_N z^{-N}).$$

If the maximum signal magnitude X_{max} does not satisfy (5.39), then the input can be multiplied by a scale factor G where

$$G = \frac{1}{X_{max}} \frac{1}{\left(\max_k \sum_{m=-\infty}^{\infty} |f_k[m]| \right)}. \qquad (5.40)$$

As we shall see in the following exercises, the introduction of scaling can greatly improve the processing accuracy. However, using scaling factors that are too small can also lead to degraded performance. In such cases, the scaling factor effectively robs the input of bits by not allowing it to fully utilize the dynamic range of the quantizer, which in turn leads to additional quantization error. If a less conservative estimate for the scaling factor is used, the quantization error can be reduced. Saturation overflow may not be eliminated completely but its occurrence should be infrequent. Such a scale factor is given by

$$G = \frac{1}{X_{max}} \frac{1}{\left(\max_k \sum_{m=-\infty}^{\infty} |f_k[m]|^2 \right)^{1/2}}. \qquad (5.41)$$

Filters are often implemented as a cascade of second-order sections. In this case, it is sometimes beneficial to distribute the scaling factors among the sections. All of these points are investigated in the following exercises, which examine the effects of roundoff and saturation overflow errors on the performance of a filter in a fixed-point processing environment.

EXERCISE 5.6.1.   **Scaling in Fixed-Point Systems**

Scaling of the input $x[n]$ is an important issue in systems that use fixed-point arithmetic. If intermediate sample values at the nodes of the flow graph are allowed to exceed unity,[10] overflow will occur that can result in distortion. Therefore, $x[n]$ should be scaled so that the overall distortion is minimized.

Let the filter $H(z)$ be the causal, stable filter

$$H(z) = \frac{0.483 + 0.174z^{-2}}{1 - 0.6z^{-1}}.$$

Note that the numerator is a second-order polynomial. Use the *create file* option in **f siggen** to create a file **hn** for this filter.

For an input signal use the file **sig3**, which is provided for you in the software.

(a) First, you will examine the performance of a direct form II implementation in a fixed-point environment. Execute the command

f direct2 sig3 hn out1 out2 −1 1 512 0 30

---

[10]Implicit in this statement is the assumption that the samples are represented as binary fractions.

The file **out1** will contain the approximate ideal output and the file **out2** will contain the output from the fixed point system assuming saturation arithmetic, nine-bit quantization (or 512 levels), and a dynamic range between -1 and $+1$. Display the output signals on the screen using **g sview2** and observe that there are noticeable differences. These differences are due primarily to saturation overflow. This type of overflow distortion occurs when the intermediate sum of two numbers falls outside the -1 to $+1$ quantization range. The out-of-bound sum is set to the minimum or maximum quantization level, hence the name *saturation* distortion.

Use the **f snr** function to compute and record the signal-to-noise ratio or SNR using **out1** and **out2** as inputs. The larger the value of the SNR, the more accurate the result.

(b) Now consider applying a scale factor to the input so that the magnitude of the intermediate sums that are computed in the difference equations do not exceed unity. Use (5.40) to determine the scaling factor. To do this you must first compute the impulse response for the $f_k[n]$. Since there is just one summation node in the direct form II implementation, there is only $f_1[n]$. This filter, $f_1[n]$, is the impulse response corresponding to $1/A(z)$, where $H(z) = B(z)/A(z)$. We can compute this impulse response by first creating an IIR file for $1/A(z)$ and then explicitly computing its response to an impulse. The following commands will allow you to compute 50 samples of the impulse response:

> **f convert   hn   tmp1   tmp2**
> **f revert   impulse   tmp2   tmp3**
> **f filter   impulse   tmp3   fn   50**

The file **fn** will contain 50 samples of the impulse response, $f_1[n]$. Display and sketch $f_1[n]$ using **g view**. Equation (5.40) can be evaluated easily with the help of the **f mag** and **f summer** functions. Determine the scale factor implied by (5.40). You may assume that $X_{max} = 1$.

(c) Use the **f gain** function to multiply the input by the scaling factor to form **xn**. Evaluate the filtered output (as in part (a)) but using the scaled input, **xn**. Call the two outputs **out3** and **out4**. Compute the SNR by typing

> **f   snr   out3   out4**

and observe the improvement over the results obtained in the previous part.   □

EXERCISE 5.6.2.   More on Scaling

The previous exercise showed that scaling could eliminate saturation overflow and the distortion that results from it. Significant improvement in quality was observed as a result. However, conservative scaling estimates do not always allow for the best performance because they trade overflow distortion for increased quantization noise. For example, if the signal was originally represented in a ten-bit representation and then scaled by a factor of 2^{-9}, all the bits would be shifted 9 places to the right, leaving only one bit to represent the signal. Although this may help the saturation problem, it elevates the quantization error. Clearly, there is a tradeoff that should be made between errors due to overflow and errors due to overscaling.

If a less conservative estimate for the scaling factor is used, such as that given in (5.41), the quantization error can often be reduced with only a small increase in saturation error. Repeat the previous exercise using (5.41). How does the SNR obtained compare to that obtained using the scale factor in (5.40)?   □

EXERCISE 5.6.3.   Some Other Scaling Methods

The previous exercises illustrated the tradeoff between saturation overflow distortion and the distortion due to overscaling. Here you consider another scaling function where

$$G = \frac{1}{\max_{k,\omega}\{|F_k(e^{j\omega})|\}}.$$ (5.42)

The value of $\max_{k,\omega}\{|F_k(e^{j\omega})|\}$ is obtained by computing the DTFT and measuring the peak value of the magnitude response.

Show analytically that this scaling function is always equal to or greater than G in (5.40), but always less than or equal to G in (5.41). This scaling function can also provide useful scale factors for fixed-point systems.   □

EXERCISE 5.6.4.   Scaling in a Cascade Structure

The cascade form realization of digital filters is a popular implementation structure because of its relative insensitivity to coefficient quantization. In this exercise, we continue our examination of scaling but for cascade realizations.

The files **h0c1**, **h0c2**, **h0c3**, and **h0c4** contain the second-order sections for the bandpass filter **h0df**. These sections will used to form the cascaded implementation. Design an input signal, $x[n]$, of length 60 using **f rgen** with a seed value of 1. Store $x[n]$ in the file **xn**.

(a) First consider the output of the cascaded system in the absence quantization error. The following commands will evaluate the "ideal" output. Put these commands in a macro called **cascade1.bat** and run the macro.

```
f direct1   xn     h0c1   out1   out0   −1   1   512   0   200
f direct1   out1   h0c2   out2   out0   −1   1   512   0   200
f direct1   out2   h0c3   out3   out0   −1   1   512   0   200
f direct1   out3   h0c4   out4   out0   −1   1   512   0   100
```

The macro will place the filtered output $y[n]$ in the file **out4**.

Next, evaluate the quantized output that contains all the errors due to roundoff and saturation overflow in the fixed-point environment with nine-bit quantization. This may be done using the commands

```
f direct1   xn     h0c1   out   out1   −1   1   512   0   200
f direct1   out1   h0c2   out   out2   −1   1   512   0   200
f direct1   out2   h0c3   out   out3   −1   1   512   0   200
f direct1   out3   h0c4   out   out5   −1   1   512   0   100
```

Put these commands in a macro called **cascade2.bat** and run the macro. The output $\hat{y}[n]$ will be placed in the file **out5**. Display and sketch the two outputs **out4** and **out5** using the **g sview2** function. Compute the SNR by typing **f snr out4 out5** and notice the poor performance.

(b) The poor performance in part (a) is due largely to saturation overflow. Following the procedure outlined in Exercise 5.6.1 determine the scale factor based on (5.41). Displaying $x[n]$ will reveal that $X_{max} = 0.5$. This value should be used when evaluating (5.41).

Note that here you will have four $f_k[n]$ functions to evaluate, each of which can be evaluated using the commands given in Exercise 5.6.1b. You may use the **f gain** function to perform the input scaling. Apply your scaling factor as described, compute the quantized output $\hat{y}[n]$ using the macro **cascade2.bat**. Display the quantized and unquantized outputs using **g sview2**. Compute and record the SNR by typing **f snr out4 out5**.

(c) Better results should be attainable if the scale factors are distributed among the summation nodes in the cascade. Determine four scale factors—one to precede each second-order section—that lead to further improvement. The scale factors may be applied in the following way:

```
f gain    input    input1    factor1
f direct1    input1    h0c1    out    temp    −1    1    512    0    200
f gain    temp    out1    factor2
f direct1    out1    h0c2    out    temp    −1    1    512    0    200
f gain    temp    out2    factor3
f direct1    out2    h0c3    out    temp    −1    1    512    0    200
f gain    temp    out3    factor4
f direct1    out3    h0c4    out    temp    −1    1    512    0    100
f gain    temp    out6    rfactor
```

where **factor1, factor2, factor3,** and **factor4** are the scale factors that you determined, and **rfactor** is the reciprocal of the product of all the scale factors. Notice that **rfactor** restores the output to its proper amplitude. For convenience, put these commands in a macro.

Compute the SNR using **out0** from part (a) as the ideal. This may be done by typing **f snr out0 out6**. Compare this SNR with that found in part (b). Attempt to manually tweak or adjust the scale factors to achieve the highest possible SNR. Record your best set of scale factors and the corresponding SNR.   □

5.7 LIMIT CYCLES

Other nonlinear distortions caused by quantization that often plague IIR filter implementations are limit cycles. These can be caused either by nonlinear quantization or by the overflow of additions. The problem may be described as follows. The output of a stable LTI filter is expected to decay to zero after the input to the system becomes zero.

However, because quantization makes these filters nonlinear, and this nonlinearity occurs in a feedback loop, it is possible to have sustained oscillations at the output in spite of the input being zero. This phenomenon, which is called a *zero-input limit cycle,* is now demonstrated in the exercises.

EXERCISE 5.7.1.   Limit Cycles in First-Order Systems

The simplest illustration of a limit cycle is the for the case of a causal first-order system with one pole. Such a system has the transfer function

$$H(z) = \frac{1}{1 - \alpha z^{-1}} \tag{5.43}$$

and can be implemented using the difference equation

$$y[n] = \alpha y[n-1] + x[n]. \tag{5.44}$$

In a finite precision arithmetic environment, this equation becomes nonlinear, as we have seen, and is given by

$$\hat{y}[n] = Q\{\alpha \hat{y}[n-1]\} + \hat{x}[n], \tag{5.45}$$

where $\hat{x}[n]$ is a quantized input signal. For this example, assume the input to be the unit sample $\delta[n]$ store in the file **impulse**. Use the *create file* option of **f siggen** to generate the filter in (5.43) with $\alpha = 0.4$. When creating the file for $H(z)$, note that the numerator order is zero, the denominator order is 1, $b_0 = 1$, and $a_1 = -0.4$.

(a) Use **f direct1** to compute the impulse response of this filter. Specify -1 and $+1$ as the minimum and maximum quantizer values, 31 as the number of levels, 0 for saturation arithmetic, and 30 for the number of output samples. The function will output two files: the first being the result with approximately infinite precision (unquantized) arithmetic and the second with finite precision (quantized) arithmetic as specified above. Using **g view2**, display and sketch the two outputs. You may notice small effects due to quantization. However, the effects are small and both plots should display the same general profile.

(b) Use a text editor to change the value of α in the filter to 0.7 or you may simply redesign the filter with the new value of α. Repeat the evaluation as described in part (a) and sketch the two output plots. Unlike the unquantized case, the output in the quantized case locks up after a certain point and does not decay to zero as it should. This is a zero-input limit cycle. The amplitude below which no further decay occurs is called the *dead band.*

(c) Repeat part (a) for $\alpha = 0.8$ and $\alpha = 0.9$. Record the value of the dead band in each case. What appears to be the relationship between α and the dead band?   □

EXERCISE 5.7.2.   More Limit Cycles in First-Order Systems

(a) Repeat part (a) in Exercise 5.7.1 using $\alpha = -0.7$. Notice that when creating this file using the *create file* option in **f siggen** the numerator order is zero, the denominator order is 1, $b_0 = 1$, and $a_1 = 0.7$. You will observe that the output, after a certain point, oscillates. What is the amplitude range of this oscillation? For limit cycles of this type, the peak-to-peak amplitude range is the dead band. The oscillation behavior seen in the output is the motivation for the name *periodic limit cycle*. In this case, the period is two. For the previous exercise, the period is one.

(b) Now repeat the evaluation with 17 and 11 quantization levels. How does the number of quantization levels affect the dead band? ☐

EXERCISE 5.7.3.   **Limit Cycles in First-Order Systems**

The analysis of limit cycles in general is quite difficult; however, for first- and second-order systems, useful interpretations can be given. Jackson [5] advanced the viewpoint that a system with a limit cycle behaves like a system with poles on the unit circle. Consider the first-order system

$$H(z) = \frac{1}{1 - \alpha z^{-1}}$$

with the corresponding nonlinear difference equation

$$\hat{y}[n] = Q\{\alpha\hat{y}[n-1]\} + \hat{x}[n].$$

During the time period when the system is in the limit cycle

$$Q\{\alpha\hat{y}[n-1]\} = \text{sgn}[\alpha]\hat{y}[n-1], \tag{5.46}$$

where

$$\text{sgn}[\alpha] = \begin{cases} 1, & \text{if} \quad \alpha \geq 0 \\ -1, & \text{if} \quad \alpha < 0 \end{cases}. \tag{5.47}$$

After substitution into (5.45), we get the equation

$$\hat{y}[n] = \text{sgn}[\alpha]\hat{y}[n-1] + \hat{x}[n].$$

This is a system with a pole on the unit circle at either $z = 1$ or $z = -1$ depending on the sign of α. To determine the actual bounds of the dead band, recall that rounding implies that

$$|Q[\alpha\hat{y}[n-1]] - \alpha\hat{y}[n-1]| \leq \frac{1}{2}\Delta,$$

where Δ is the quantizer step size. This combined with (5.47) implies that

$$|\text{sign}[\alpha]\hat{y}[n-1] - \alpha\hat{y}[n-1]| = |\hat{y}[n-1] - |\alpha|\hat{y}[n-1]| \leq \frac{1}{2}\Delta,$$

which results in

$$|\hat{y}[n-1]| \leq \frac{\frac{1}{2}\Delta}{1 - |\alpha|}.$$

This inequality specifies upper bounds on the dead band.

(a) Using the inequality above, predict the dead band for the following cases:

(i) $\alpha = 2/3,$

(ii) $\alpha = 4/5,$

where $X_{min} = -1$, $X_{max} = 1$, and 21 quantization levels are assumed.

(b) Following the procedure in Exercise 5.7.1a, use **f direct1** with saturation arithmetic to explicitly compute the impulse responses for cases (i) and (ii) in part (a). Remember, when creating the filter, that $\alpha = 2/3, 4/5$ implies that $a_1 = -2/3, -4/5$, respectively.

Display and sketch the outputs using **g view2**. How well do the experimental results agree with the predictions?   □

EXERCISE 5.7.4.   **Numerical Examination of Limit Cycles**

To obtain a better understanding of why quantization error causes limit cycles, consider the simple first-order system

$$H(z) = \frac{1}{1 - 0.6z^{-1}}. \tag{5.48}$$

The difference equation corresponding to this system is

$$y[n] = 0.6y[n-1] + x[n]. \tag{5.49}$$

Letting the input be $x[n] = \delta[n]$, evaluate the difference equation manually in table form as shown

n	$x[n]$	$y[n-1]$	$y[n]$
$n = -1$	$x[-1] = 0$	$y[-2] = 0$	$y[-1] = 0$
$n = 0$	$x[0] = 1$	$y[-1] = 0$	$y[0] = 1$
$n = 1$	$x[1] = 0$	$y[0] = 1$	$y[1] = .6$
$n = 2$	$x[2] = 0$	$y[1] = .6$	$y[2] = .36$
$n = 3$	$x[3] = 0$	$y[2] = .36$	$y[3] = .216$
⋮	⋮	⋮	⋮

Complete the chart for $n = 4$ through $n = 8$.

Next use the *create file* option of that function to generate the filter in (5.49). Use **f direct1** and the file **impulse** to compute the impulse response of this filter with minimum and maximum quantizer values of -1 and $+1$, 31 levels, 0 for saturation arithmetic and 0 for the number of output samples. This last parameter will execute the function in its verbose mode where intermediate values are displayed on the screen. Step through the function by pressing the "c" key and record these quantized intermediate values in a chart similar to the one above. In particular, record the values for $x[n]$ (the second column of numbers) and $\hat{y}[n]$, which appears as $qy[\cdot]$ on the screen. Compare the two charts. What is the cause of the limit cycle behavior?   □

EXERCISE 5.7.5. **Limit Cycles in Second-Order Systems**

Consider the causal, stable second-order system

$$y[n] = x[n] + \alpha_1 y[n-1] + \alpha_2 y[n-2]. \qquad (5.50)$$

The finite-register-length implementation is given by

$$\hat{y}[n] = x[n] + Q[\alpha_1 \hat{y}[n-1]] + Q[\alpha_2 \hat{y}[n-2]], \qquad (5.51)$$

where we assume $X_{min} = -1$, $X_{max} = 1$, and the number of quantization levels equals 17. An interesting mode of limit cycle behavior occurs when roundoff noise forces the equation to behave like

$$\hat{y}[n] = x[n] + Q[\alpha_1 \hat{y}[n-1]] - y[n-2].$$

Here we have apparent complex poles on the unit circle. In this case the dead band is bounded by

$$|y[n]| \leq \frac{\frac{1}{2}\Delta}{|1+\alpha_2|}.$$

Use the *create file* option of **f siggen** to generate three filters in the form of (5.50) with $\alpha_1 = 0$, 0.5, 1.5, and $\alpha_2 = -0.9$. Note that this means $a_1 = 0$, -0.5, -1.5, respectively, and $a_2 = 0.9$.

Use **f direct1** to compute the impulse response of the filter using the file **impulse** as input and with minimum and maximum quantizer values of -1 and $+1$, 31 as the number of levels, 0 for saturation arithmetic, and 100 for the number of output samples.

What do you observe about the general relationship between α_1 and the period of oscillation in the limit cycle?   □

EXERCISE 5.7.6. **Limit Cycles Due to Overflows**

Filtering in processing environments that use finite precision arithmetic with wraparound overflow can lead to another form of limit cycle behavior. Limit cycles of this type can produce large output oscillations that can occupy a major part of the quantization range of the system. To illustrate such limit cycles, consider the system

$$\hat{y}[n] = Q_c\{x[n] + Q_c[\alpha_1 \hat{y}[n-1]] + Q_c[\alpha_1 \hat{y}[n-2]]\}$$

with a unit sample input. Use the *create file* option to design the filter

$$H(z) = \frac{1}{1 - 0.7807z^{-1} + 0.7807z^{-2}}.$$

Notice that the numerator order should be 0, the denominator order 2, the starting point 0, $a_1 = -0.7807$, and $a_2 = +0.7807$.

(a) Using **g polezero**, display and sketch the pole/zero and magnitude response plots for $H(z)$.

(b) Use the file **impulse** to examine the impulse response of $H(z)$, using the **f direct1** function with $X_{min} = -1$, $X_{max} = +1$, number of levels equal to 39, wraparound overflow, and 80 output samples. This information may be specified directly on the command line by typing

f direct1 impulse $H(z)$ **out1 out2** -1 **1 39 1 80**

where **out2** is the 80-sample impulse response based on a 39-level number representation and **out1** is the same but based on a floating-point representation. Use **g view2** to display the files **out1** and **out2** and sketch these results. Are the differences in the signals due to quantization effects visually noticeable?

(c) Create an input file called **input** that represents the signal

$$x[n] = 0.7837\delta[n] + 0.6084\delta[n-1],$$

using the *create file* option in **f siggen**. Again, evaluate 80 samples of the output response using **f direct1** with the same specifications as before, i.e., type

f direct1 input $H(z)$ **out1 out2** -1 **1 39 1 80**

Display **out1** and **out2** using **g view2** and sketch the results. The signal in file **out1** represents the infinite precision output. Note the severity of the limit cycle behavior shown here and that there is virtually no resemblance between the floating-point and fixed-point outputs.

(d) Explain why the catastrophic limit cycle behavior in the previous part occurred. To do this, examine the filtering process one step at a time in table form. Start with the table

n	$y[n]$	$x[n]$	$y[n-1]$	$y[n-2]$
$n = 0$	0.7837	0.7837	0.00	0.00
$\vdots$	$\vdots$	$\vdots$	$\vdots$	$\vdots$

and manually complete the evaluation through $n = 3$. You should verify that your manual calculations agree with the computer generated output computed earlier. Now evaluate the **f direct1** function in verbose mode. This can be done by typing

f direct1 input $H(z)$ **out1 out2** -1 **1 39 1 0**

The function, operating in this mode, will take you through the computation of the difference equation step by step and show you the intermediate values that result when the products and sums in the difference equation are quantized with wraparound overflow. Note that the quantized output $qy[\cdot]$ displayed on the screen is $b_0 x[n] - a_1\hat{y}[n] - a_2\hat{y}[n-1]$ and that $qy[\cdot]$ reflects the effects of wraparound quantization. Compare the quantized results with the unquantized results you computed in your chart. Identify where the major errors occur and explain why the output error is so large.

(e) Will this same type of limit cycle behavior occur if saturation arithmetic is used? Try evaluating 80 samples of the difference equation using saturation arithmetic instead of wraparound. This can be done by typing

f direct1   input   H(z)   out1   out2   −1   1   39   0   80

(f) Will this type of limit cycle behavior exist in a direct form II implementation with wraparound overflow? What about with saturation overflow? Try them using the **f direct2** function instead of **f direct1**.   □

EXERCISE 5.7.7.   More on Limit Cycles Due to Overflows

Carefully work the previous exercise if you have not already done so and pay particular attention to the cause of this type of limit cycle.

Determine another second-order filter $G(z)$, input signal $x[n]$, and number of quantization levels that produce the same kind of catastrophic limit cycle behavior observed in part (c) of the previous exercise. Record these values.

Display the pole/zero and magnitude response plots for $G(z)$ using **g polezero** and sketch them. Your filter, $G(z)$, must be causal and stable, i.e., the poles must be inside the unit circle.

Evaluate 80 samples of the difference equation using **f direct1** and display and sketch the resulting outputs based on fixed-point and floating-point arithmetic that illustrate this type of limit cycle behavior.   □

REFERENCES

[1] A. H. Gray and J. D. Markel, "Digital lattice and ladder filter synthesis," *IEEE Trans. Acoustics, Speech, and Signal Proc.*, Vol. 21 (Dec. 1973), pp. 491–500.

[2] A. H. Gray and J. D. Markel, "A normalized digital filter structure," *IEEE Trans. Acoustics, Speech, and Signal Proc.*, Vol. 23 (June 1975), pp. 258–277.

[3] J. F. Kaiser, "Digital filters," Chapter 7 in *System Analysis by Digital Computer*, F. F. Kuo and J. F. Kaiser, eds., John Wiley, New York (1966).

[4] C. Rader and B. Gold, "Effects of parameter quantization on the poles of a digital filter," *Proc. IEEE* (May 1967), Vol. 55, pp. 688–689.

[5] L. Jackson, "An analysis of limit cycles due to multiplicative rounding in recursive digital filters," *Proc. 7th Allerton Conf. Circuit System Theory* (1969), pp. 69–78.

Design Projects

6

The previous chapters provided hands-on exposure to the basic principles of digital filter design. With this as a foundation we next examine a more complex set of problems in the form of several design projects. You may wish to reference some of the introductory discussions in the earlier chapters as you work through them. The projects included here require careful thought and allow you to be creative in formulating solutions.

Some projects have an analytical flavor while others are more experimental. Some may be performed exclusively using the DSP software provided on the disk, but several others require that you write computer programs in a language of your choice, for which you will need a compiler and a working knowledge of the program language. You may find the discussion at the end of the appendix useful. It describes the DSP format in sufficient detail to allow you to read and write files in the DSP file format.

6.1 NUTTALL–BESSEL FILTER DESIGN

The Kaiser window that you explored in Section 2.2 is particularly useful for designing digital filters because it has a control parameter α that allows you to design a wide variety of filters with different transition band widths and passband/stopband deviations. Although this window leads to a very convenient filter design procedure, we note that the technique does not provide control over the shape of the rolloff of the stopband ripples. We observed in Chapter 2 that the ripples in the stopband of a lowpass filter designed using a Kaiser window decrease monotonically in magnitude as they move away from the cutoff frequency. In the project, you will evaluate a new window for filter design that provides some control over the slope of this stopband rolloff.

Kaiser [1] has suggested that the Nuttall–Bessel window [2] would provide such a capability if integrated into a window-based filter design procedure. In this project, you will investigate the Nuttall–Bessel window and write a user-friendly computer program for FIR filter design that uses it.

The normalized Nuttall–Bessel window that forms the basis for the design is given by

$$w[n] = \left(\sqrt{1 - (n/N)^2}\right)^\nu \frac{I_\nu(\alpha\sqrt{1 - (n/N)^2})}{I_\nu(\alpha)}, \quad |n| \leq N,$$

where α and ν are two parameters that define the shape of the window and the window length is $L = 2N + 1$, with $-N \leq n \leq N$.

Several points are important to mention. First, $I_\nu(\cdot)$ is the modified Bessel function of the first kind of order ν. (Note that ν is not necessarily an integer.) When $\nu = 0$, the window reduces to the Kaiser window. Thus the role of the parameter α in the Nuttall–Bessel window is similar to its role with the Kaiser window. Second, the modified Bessel function is defined as

$$I_\nu(x) = \sum_{k=0}^{\infty} \frac{1}{k!\Gamma(\nu + k + 1)}(\frac{x}{2})^{\nu+2k},$$

where $\Gamma(\cdot)$ is the gamma function. The gamma function, in turn, is defined by

$$\Gamma(x) = \int_0^{\infty} t^{x-1}e^{-t}\,dt.$$

It can be generated from the functional equations

$$\Gamma(k + x) = (x + k - 1)(x + k - 2)\cdots(x + 1)\Gamma(x + 1),$$

$$\Gamma(k + x) = \frac{\Gamma(x + 1)}{(x - k)(x - k + 1)(x - k + 2)\cdots(x - 1)x},$$

where x is a real number and k is a nonnegative integer. For $0 \leq x \leq 1$,

$$\Gamma(x + 1) \approx 1 + a_1 x + a_2 x^2 + a_3 x^3 + a_4 x^4 + a_5 x^5,$$

where

$$a_1 = -0.5748646,$$
$$a_2 = 0.9512363,$$
$$a_3 = -0.6998588,$$
$$a_4 = 0.4245549,$$
$$a_5 = -0.1010678.$$

The gamma function has a couple of useful simplified forms for special cases of its argument:

$$\Gamma(n + 1) = n! \quad n = 0, 1, 2, \ldots \tag{6.1}$$

$$\Gamma(n + 1/2) = \frac{1 \cdot 3 \cdot 5 \cdots (2n - 1)}{2^n}\sqrt{\pi} \quad n = 0, 1, 2, \ldots \tag{6.2}$$

$$\Gamma(-n + 1/2) = \frac{(-1)^n 2^n \sqrt{\pi}}{1 \cdot 3 \cdot 5 \cdots (2n - 1)} \quad n = 1, 2, 3, \ldots \tag{6.3}$$

The gamma function $\Gamma(x)$ is defined for negative values of x, but singularities exist when x is a negative integer.

Finally, both ν and α have a clear impact on the shape of the Nuttall–Bessel window. However, the relationship between the ν and α parameters and the stopband attenuation and stopband rolloff of a filter designed using the window is the topic for investigation.

(a) Write a computer program that designs odd length lowpass FIR filters using the Nuttall–Bessel window. The inputs to your program should be the following parameters: the filter length, the values of ν and α for the window, and the nominal cutoff frequency of the filter. The output should be a filter file in the standard DSP file format so that the filter can be examined using the DSP software. See the end of the appendix for a discussion on writing files in the DSP file format.

(b) Your objective is to determine a relationship between the window parameters ν and α and the filter design parameters δ_s, δ_p, ω_p, ω_s, the slope of the stopband rolloff, and the filter length. After obtaining an empirical relationship among these parameters, integrate it into the filter design program. The resulting program should allow the user to specify the ripple amplitude, the cutoff frequencies, the rolloff, and the filter length. Try to make the design program as accurate as possible for designing filters that are approximately equiripple, i.e., zero rolloff.   □

6.2 WEIGHTED LEAST-SQUARES FIR DESIGN

The Parks–McClellan algorithm designs FIR filters that are optimal in the Chebyshev or minimax sense. In this project, we consider the design of filters that are optimal in terms of minimizing a weighted squared error. The filters we will explore are based on designs described in [3]. Another form of these designs was later discovered by Vaidyanathan and Nguyen [4] and given the name *eigenfilters*.

The easiest way to describe the technique is to first consider only the design of Type I FIR filters (odd-length filters with symmetric impulse responses). Since the methodology can work with an arbitrary positive weighting function and ideal frequency response, the technique can be straightforwardly modified for Type II, Type III, and Type IV filters using the same methods that were described for the Parks–McClellan designs in Exercises 2.4.2–2.4.4.

Let us begin by writing the Type I filter as a zero-phase filter, $H(e^{j\omega})$, by shifting the impulse response so that it corresponds to an even sequence. The frequency response of the filter can then be written in the form

$$H(e^{j\omega}) = \sum_{n=0}^{N} a[n] \cos \omega n,$$

where

$$a[n] = \begin{cases} h[n], & n = 0 \\ 2h[n], & 1 \le n \le N. \end{cases}$$

The filter length L is $2N + 1$.

The goal of the design procedure is to find the coefficients that represent the best approximation to an ideal frequency response $I(e^{j\omega})$ in a weighted least-squares sense. Mathematically, the coefficients should be chosen to minimize

$$E = \int\limits_{-\pi}^{\pi} \left| W(e^{j\omega}) \left[H(e^{j\omega}) - I(e^{j\omega}) \right] \right|^2 \, d\omega. \tag{6.4}$$

Practical considerations usually dictate that this error be replaced by a discrete version such as

$$E' = \sum_i \left| W(e^{j\omega_i}) \left[H(e^{j\omega_i}) - I(e^{j\omega_i}) \right] \right|^2, \tag{6.5}$$

where the set $\{\omega_i\}$ corresponds to samples taken along the frequency axis.

For a Type I design such as a lowpass filter, the weighting function $W(e^{j\omega})$ might take the form

$$W(e^{j\omega}) = \begin{cases} K, & 0 \le \omega \le \omega_p \\ 0, & \omega_p \le \omega \le \omega_s \\ 1, & \omega_s \le \omega \le \pi. \end{cases}$$

Solving this problem is straightforward. By calculating the gradient of E with respect to the unknowns $\{a[n]\}$ and setting the result to zero, the unknowns can be found by solving the following set of linear equations:

$$\sum_{\ell=0}^{N} a[\ell]\phi_{\ell k} = I_k, \qquad k = 0, 1, \ldots, N,$$

where

$$\phi_{k\ell} = \int\limits_{-\pi}^{\pi} W^2(e^{j\omega}) \cos(\ell\omega) \cos(k\omega) \, d\omega$$

and

$$I_k = \int\limits_{-\pi}^{\pi} W^2(e^{j\omega}) I(e^{j\omega}) \cos(k\omega) \, d\omega.$$

If the discrete problem formulation is used, the solution is similar except that the coefficients in the linear equations become

$$\phi_{k\ell} = \sum_i W^2(e^{j\omega_i}) \cos(\ell\omega_i) \cos(k\omega_i)$$

and

$$I_k = \sum_i W^2(e^{j\omega_i}) I(e^{j\omega_i}) \cos(k\omega_i).$$

The filter coefficients $a[n]$ (and in turn h[n]) may be obtained by using a subroutine for solving sets of symmetric linear equations. These may be found in a standard mathematics software library or programs book, such as [5].

Following this development, write a program in a language of your choice that designs odd length lowpass filters using a weighted mean squared error. The program should allow you to specify the passband and stopband cutoff frequencies, ω_p and ω_s, the filter length, L, and the parameter K that will give you some control over the relative heights of the passband and stopband ripples. The output should be a file containing the filter coefficients in DSP format. You may use either the integrated error (6.4) or the discretized error (6.5). In the former case, you will need to evaluate the appropriate integrals. In the latter case, your program will be longer and will probably run slower.

Empirically determine an approximate functional relationship between the weight K and the passband and stopband ripples δ_p and δ_s. Incorporate this relationship into your program so that the user also has the option to specify the stopband and passband ripples in lieu of K. □

6.3 QUANTIZATION OF OVERSAMPLED SIGNALS

With a classical A/D converter, a signal is first filtered with an analog antialiasing filter with cutoff frequency equal to the signal bandwidth Ω_c. The filtered signal is then sampled at a rate of Ω_c/π samples per second (or more) and quantized to accommodate the data wordlength of the digital system.

High quality analog antialiasing filters, unlike digital filters, tend to be expensive. This has motivated the use of oversampling A/D converters in which the signal is sampled at a rate $M\Omega_c/\pi$. The antialiasing filtering is then done digitally and the sampling rate is adjusted using an M-to-1 decimator. An analog antialiasing filter may also be employed in such systems, but because of the oversampling, its specifications can be much less stringent and hence is much easier to build. Of course, sampling at a rate of $M\Omega_c/\pi$ implies that the number of samples involved is increased by a factor of M. The digital portion of the oversampling A/D is illustrated in Figure 6.1. The signal $s[n]$ is assumed to be the oversampled analog signal, which is then decimated in the digital domain as shown in Figure 6.1a. To offset the storage and computational demands associated with the high sampling rate, the data wordlengths are typically very small. This is very important to the cost effectiveness of the system. In fact, one-bit wordlengths are not uncommon in such systems. The short data wordlength issue is illustrated in Figure 6.1b, where the quantizer indicates that the discrete signal representation at the high data rate contains a small number of bits/sample. After filtering with the lowpass filter, the data wordlength is increased to B bits (which typically lies in the 10–18 range).

Comparing the signal outputs $x[n]$ in Figure 6.1a and $x_1[n]$ in Figure 6.1b, it is evident that the two will not be identical due to quantization. In this project, you will investigate developing a quantization noise reduction system (illustrated in Figure 6.1c) that will allow $x_2[n]$ to better approximate $x[n]$.

Use **f rgen** to design a 1024-point random signal with an initial seed value of 1. Filter this signal using **f convolve** with a 64-point FIR lowpass filter with a cutoff frequency of 0.65 rad, designed using a Hamming window. This filter can be designed using the **f fdesign** function. Call the resulting signal file **sn**, which represents $s[n]$ in Figure 6.1. Use **f dnsample** to downsample **sn** by a factor of 4 and **f truncate** to truncate the

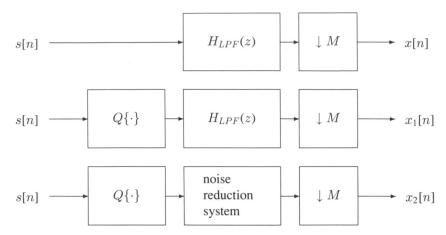

Figure 6.1. Oversampling Systems. a) Without signal quantization. b) With signal quantization. c) With quantization and noise reduction system (delta-sigma modulation).

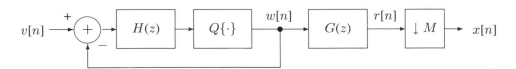

Figure 6.2. Block diagram of a delta-sigma modulator and compensation filter.

signal to $L = 255$ samples to form $x[n]$. Put the output in the file **xn**, which will be the reference signal assuming no quantization.

To simulate the quantization of the high rate signal from the oversampling A/D, use **f quantize** to quantize **sn** to 17 levels with quantization range ± 1 to form **vn**. Now downsample **vn** 4-to-1 and truncate it to 255 samples as before. Call the output **xn1**. Display **xn** and **xn1** using **g sview2**.

To measure how accurately the quantized signal **xn1** approximates **xn**, compute the SNR in the frequency domain in the following way. First pad each signal with zeros for a total length of 511 using **f zeropad**. Then compute the FFT of **xn** and **xn1** using **f fft** and take their magnitudes using **f mag**. Finally, use **f snr** to compute the SNR using the Fourier magnitude signals as inputs.

The object of this project is to develop a noise reduction system that will allow you to quantize **sn** to 17 levels and downsample the result to obtain the highest SNR possible. An approach to this problem is based on the principle used in delta-sigma modulation [6], a block diagram of which is shown in Figure 6.2. The system with input $v[n]$ and output $w[n]$ is the delta-sigma modulator. The filter $G(z)$ is a lowpass compensation filter.

(a) Show that when the quantizer in the delta-sigma modulator is removed,

$$\frac{W(z)}{V(z)} = \frac{H(z)}{1 + H(z)}.$$

(b) Now assume a linear noise source model for quantization as discussed in Chapter 5, where the additive quantization noise is $e[n]$. Show that the response due to the noise source $e[n]$ is

$$\frac{W(z)}{E(z)} = \frac{1}{1 + H(z)}.$$

For typical delta-sigma modulators, $H(z) = 1/(1 - z^{-1})$.

It is evident that distortion is introduced. Determine the compensation filter, $G(z)$, which should be applied to the output to remove this distortion. Keep in mind that $G(z)$ must also serve as an antialiasing filter for the M-to-1 downsampler. Show analytically how the delta-sigma modulator and compensation filter reduce the quantization noise in the downsampled signal. What is the effect of $G(z)$ on the quantization noise?

(c) Write a program in a language of your choice that implements the delta-sigma modulation system and the compensation filter shown in Figure 6.2. The input to the system should be the oversampled signal **vn**. The output should be $r_2[n]$ with signal file **rn2**, in DSP format, as illustrated in Figure 6.1c. Downsample **rn2** 4-to-1 using **f dnsample** to obtain **xn2**. Compute the SNR in the frequency domain as you did before using the FFT and **f snr**. Your output file **xn2** may be longer than 256 samples due to the inherent filtering involved. This implies that a system delay has been introduced. However, the frequency domain SNR measure outlined in this project is not affected by this delay, which is the motivation for its use. The delta-sigma modulation based system should show an improvement in the SNR.

(d) Using this delta-sigma modulation principle and your own creativity, modify your program to further increase the SNR. Note that $H(z)$ can be better designed to improve the performance. Attempt to make the SNR as high as you can.   □

6.4 IDEALS FOR WINDOW DESIGNS

Most discussions of the window method for the design of FIR filters concentrate on the selection of the window function. They ignore the design freedom inherent in the selection of the ideal prototype filter. Figure 6.3 illustrates the frequency responses for three possible lowpass prototypes that could be used to design an FIR lowpass filter using the window method. All three of these ideal frequency responses have a passband cutoff frequency of ω_p and a stopband cutoff frequency of ω_s. They all have unity gain in the passband and zero gain in the stopband; the only differences among them lie in their transition behavior. The prototype $I_1(e^{j\omega})$ is the usual choice when using the window method; it has a discontinuity at $(\omega_p + \omega_s)/2$. The other prototypes $I_2(e^{j\omega})$ and $I_3(e^{j\omega})$ remove this discontinuity by prescribing the transition response over an interval of length

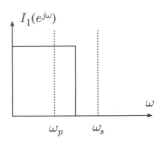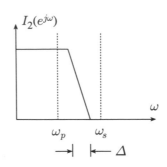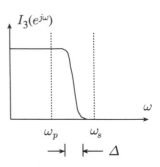

Figure 6.3. Three possible choices for the ideal frequency response of a lowpass filter to be approximated using windows.

Δ in the center of the transition band. For $I_2(e^{j\omega})$ this transition is a linear taper. This ideal response can be written in the form

$$I_2(e^{j\omega}) = I_1(e^{j\omega}) * G_2(e^{j\omega}),$$

where

$$G_2(e^{j\omega}) = \begin{cases} \Delta, & |\omega| < \Delta/2 \\ 0, & \text{otherwise.} \end{cases}$$

The other prototype, $I_3(e^{j\omega})$, which has a sinusoidal transition, can be written in a similar form

$$I_3(e^{j\omega}) = I_1(e^{j\omega}) * G_3(e^{j\omega}),$$

where

$$G_3(e^{j\omega}) = \begin{cases} \frac{\Delta}{\pi}\sin(\frac{2\pi}{\Delta}\omega), & |\omega| < \Delta/2 \\ 0, & \text{otherwise.} \end{cases}$$

Clearly the choice of the ideal that is used can have an effect on the resulting filter.

The primary goal of this project is to determine which of the three choices for the ideal frequency response results in the best lowpass filter when used with the window design method. A secondary goal is to learn how to select values for the parameter Δ for given values of ω_p and ω_s.

(a) Determine the ideal impulse responses $i_1[n]$, $i_2[n]$, and $i_3[n]$ corresponding to the three ideal frequency characteristics. For the latter two cases your results should be functions of the parameter Δ.

(b) Next you will need to develop a procedure for producing arrays of values of the sequences $i_2[n]$ and $i_3[n]$ for arbitrary values of ω_p, ω_s, and Δ. One possibility is to write a program in a language that is supported on your computer. To be readable by the DSP functions, the output file should contain the header and be in the format described in the appendix. An alternative approach is to build these prototypes using the DSP functions. In addition to the functions that you use to realize $f[n]$ and $g[n]$ you will probably make use of the functions **f ideallp** and **f multiply**.

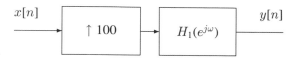

Figure 6.4.   One-stage interpolator.

(i) Design a 25-point lowpass filter using the Kaiser window method with $i_1[n]$ for $\omega_p = 0.4\pi = 1.257$ and $\omega_s = 0.6\pi = 1.885$. You should use the functions **f eformulas** (option 3) and **f kalpha** (option 3) to determine the value of the Kaiser window parameter, α, that will minimize the ripples in the passband and stopband. Examine the magnitude response of your filter using **g dfilspec** and record the maximum ripple in your design.

(ii) Design a similar 25-point filter using $i_2[n]$ for the same values of ω_p and ω_s with $\Delta = (\omega_s - \omega_p)/2$. Find the value of α by trial and error that again yields the smallest passband and stopband ripple. How do these ripples compare with the values obtained in (i)?

(iii) Repeat the design in (ii) for the ideal $i_3[n]$.

(c) For the design in either (b)(ii) or (b)(iii) above, vary the value of the parameter Δ and again optimize your design by finding the best choice for the Kaiser window parameter α. Experimentally determine the optimum values for these parameters and compare the resulting maximum ripple with the best design using $i_1[n]$ and with the equiripple design obtained using the Parks–McClellan algorithm. For the latter you should either design a 25-point FIR filter using **f pksmcc** to satisfy the same set of specifications or use **f eformulas** (option 3) to estimate the Parks–McClellan ripple.

(d) The designs so far have been performed for a specific value of N and for specific values of ω_p and ω_s. To get a feeling for how well these results generalize, repeat (c) twice—once for a filter of length $N = 41$ with the original values of ω_p and ω_s, and once for a filter of length $N = 41$ with $\omega_p = 0.45\pi = 1.414$ and $\omega_s = 0.55\pi = 1.728$.

(e) Carefully summarize your conclusions and suggest other experiments that might be run to confirm them.   □

6.5   MULTISTAGE INTERPOLATOR

In this project, you will study the practical design and implementation of a system for interpolation, such as might be required in a TDM-to-FDM[1] converter. Assume that our application requires interpolating a set of signals by a factor of $L = 100$. Two possible solutions are shown in Figures 6.4 and 6.5.   The boxes labeled $H_1(e^{j\omega})$ and $H_2(e^{j\omega})$

---

[1]TDMs and FDMs are time-division and frequency-division multiplexers. TDM and FDM approaches are two popular methods for transmitting multiple signals over a single channel.

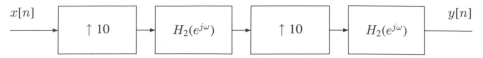

Figure 6.5.   Two-stage interpolator.

are lowpass filters; the boxes labeled $\uparrow L$ are upsamplers. Upsamplers were reviewed in Chapter 4. When $x[n]$ and $w[n]$ denote the input and output of an upsampler, then

$$w[n] = \begin{cases} x[n/L], & \text{if } n = 0, \ \pm L, \ \pm 2L, \ 3L, \ldots. \\ 0, & \text{otherwise.} \end{cases}$$

The upsampled signal is sampled at a sampling rate that is L times that of the input signal.

(a) If the ideal lowpass filters $H_1(e^{j\omega})$ and $H_2(e^{j\omega})$ in Figure 6.4 are appropriately chosen, the configurations in Figures 6.3 and 6.4 will produce the same output. Determine and plot the frequency responses of these *ideal* filters.

(b) Now suppose that the input sequence $x[n]$ was obtained by ideally sampling a continuous-time signal $x_c(t)$, i.e., $x[n] = x_c(nT)$. Further assume that the Fourier transform of the continuous-time signal is bandlimited so that $X_c(\Omega) = 0$ for $|\Omega| \geq \Omega_N$ and that the sampling period T is chosen so that $\Omega_N = (0.9)\pi/T$. Sketch the spectrum of a "typical" input signal, $X_c(\Omega)$, and the discrete-time Fourier transform of the sequence of samples, $X(e^{j\omega})$.

(c) For this part of the project you are to complete the design of the two 1:100 interpolators given in Figures 6.4 and 6.5. The errors introduced by all of the filtering and upsampling operations should be less than 0.01 at all frequencies that were less than $0.9\pi/T$ or greater than $1.1\pi/T$ in the original signal. The filters should be designed so that the overall system requires the minimum number of multiplications. The following hints and suggestions are offered for consideration.

   (i) Use the signal model in (b) to determine how to set the cutoff frequencies. The transition width should be as wide as possible to minimize the filter lengths. For the two-stage design, you should not assume that the two filters will be identical.

   (ii) The constraints on the passband and stopband ripple are for the overall system. These partially constrain the ripples of the individual filters for the two-stage system, but you nonetheless have some design freedom that you should use to minimize the filter order.

   (iii) Design your filters using either the Kaiser window method (using the functions **f eformulas**, **f kalpha**, **f kaiser**, **f ideallp**, and **f multiply**) or the Parks–McClellan algorithm using the function **f pksmcc**. Again, however, remember that the goal is to minimize the total number of multiplications while satisfying the design constraints. In counting multiplications, you should take into account the fact that for each filter the majority of its input samples are zero.

(iv) Some of the filter design programs may have a built-in gain of unity. If this is the case, make sure that the gains of your filters are adjusted accordingly.

(v) Verify the operation of your overall systems. Sketch its overall frequency response, the specifications of the filters that you used, and the total number of multiplications required for the two implementations. To determine the overall impulse responses, you can apply an impulse as input to the system using the functions **f filter** and **f upsample**. You can then calculate the effective frequency response using the function **g dfilspec**. Be aware, however, that **g dfilspec** is limited to inputs of length 512 or less.

(d) The two-stage filter that you designed above is not the best two-stage filter for this task. Better designs should be possible by changing the upsampling ratios in the two upsamplers (which will require corresponding changes in the two filters). If the two ratios are (L_1, L_2), then other possible choices include $(50, 2)$, $(25, 4)$, $(4, 25)$, and $(2, 50)$. By considering other possible choices, design the best two-stage 1:100 interpolator that you can. Notice that when the upsampling ratio is two, the filters can be designed so that nearly half of the coefficients are zero, if the filter length is odd. This fact can be exploited when implementing the filters and should be taken into account when counting the number of multiplications. Record the same information that you did for the $(10, 10)$ design.

(e) You should have observed that the best two-stage design is considerably better than the best one-stage design. For a 1:100 interpolator, however, up to four stages can be used. See if you can design a three- or four-stage interpolator that requires fewer multiplications than your best two-stage design. Record the complete design specifications and multiplication count of your best effort.   □

6.6 MULTIRATE FILTER IMPLEMENTATION

This project is concerned with the multirate implementation of a digital lowpass filter. You saw how multirate implementations could save significant computations in Chapter 4. This project is concerned with the filter design issues associated with the *design* of multirate systems.

(a) Consider the cascade of a decimator with a decimation ratio of $M = 4$ with an interpolator with an interpolation ratio of $L = 4$ as shown in Figure 6.6.

(i) What is the equivalent frequency response of the cascade?

(ii) What is the minimum order elliptic filter that will approximate the equivalent frequency response? Use

$$\delta_1 = \delta_2 = 0.01$$
$$\omega_p = 2\pi; \qquad \omega_s = 0.3\pi.$$

How many multiplications per output sample would have to be performed to realize this filter, if the filter were realized as a cascade of second-order

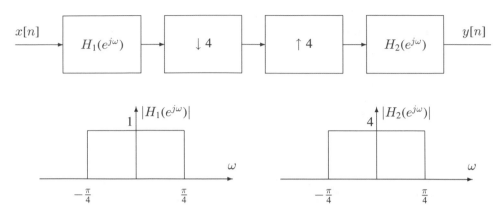

Figure 6.6. A cascade of a decimator and an interpolator.

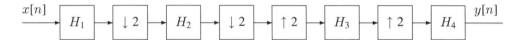

Figure 6.7. A two-stage decimator cascaded with a two-stage interpolator.

sections? What if the filter were realized as a parallel combination of two allpass filters?

(iii) Determine the minimum order FIR filter that will meet the specifications using a Kaiser window design. It is not necessary to design the filter. Its order can be estimated using the function **f eformulas**. How many multiplications are required to implement the filter?

(iv) Repeat (iii) for a linear-phase equiripple FIR lowpass filter designed using the Parks–McClellan algorithm. Again it is not necessary to actually design the filter. The function **f eformulas** may be used to estimate the order.

(b) The previous system was somewhat trivial. Replace the decimator and interpolator with two-stage versions resulting in the system shown in Figure 6.7.

(i) Sketch frequency responses for the ideal filters $H_1(e^{j\omega})$, $H_2(e^{j\omega})$, $H_3(e^{j\omega})$, and $H_4(e^{j\omega})$ that will make the cascade perform as before.

(ii) Now we must actually design the filters. Determine values for the passband and stopband cutoff frequencies and for the passband and stopband ripples that will produce filters of minimum order. Then design four filters to serve as initial candidates for the four filters in the implementation using the Kaiser window method.

(c) The overall filter complexity is controlled by four factors:

(i) Only half of the outputs of $h_1[n]$ and $h_2[n]$ are used;

(ii) Half of the inputs to $h_3[n]$ and $h_4[n]$ are zero;

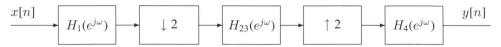

Figure 6.8. Simplification of the previous system.

(iii) $h_2[n]$ and $h_3[n]$ operate at only half the sampling rate of the other two filters;

(iv) For any filter for which $\omega_p + \omega_2 = \pi$ and $\delta_p = \delta_s$, nearly half the filter coefficients are identically zero.

Given these facts, how many multiplications per output sample are required? Compare your answer with the results in part (a).

(d) By varying the various design parameters of the four filters, produce a design that requires the minimum number of multiplications and record your best design.

(e) Repeat the whole procedure using the Parks-McClellan algorithm. Does this produce a more efficient design?

(f) An alternative realization for the filter is shown in Figure 6.8 that approximates the system in Figure 6.6. Repeat the complete design procedure for this implementation.

(g) Compare the various realizations that you considered in this project with respect to the total number of required computations and required coefficient storage.   □

6.7   OCTAVE-BAND SPECTRUM ANALYZER

The purpose of this project[2] is to design an "octave-band" spectrum analyzer (OBSA)—a system that measures power versus frequency over frequency bands whose widths have a constant ratio of 2.

For this specific problem, assume that the signal has been ideally sampled (i.e., no aliasing, quantization, etc.) at a rate of 40 kHz. You are to investigate several different designs for a 4-channel spectrum analyzer with frequency coverage as shown in the table below.

Channel Number	Lower Bandedge	Upper Bandedge
1	1 kHz	2 kHz
2	2 kHz	4 kHz
3	4 kHz	8 kHz
4	8 kHz	16 kHz

Since it is not possible to separate each channel exactly at its boundaries, we must compromise by allowing an error over the in-band frequency range. We will specify this in-band error by requiring that the response of any single channel be true to within

---

[2]This project was written by Prof. James H. McClellan.

±10% over 70% of the channel's bandwidth. The rejection of out-of-band frequencies must be at least 40 dB.

(a) The most obvious way to design this system is to provide four bandpass filters operating in parallel. Clearly list the specifications that you would use for each channel to meet the overall specifications for the OBSA and design minimum order Chebyshev IIR filters that will meet these specifications. Summarize the characteristics of each channel bandpass filter in a table that contains entries for:

  (i) The specified cutoff frequencies for passbands and stopbands,

  (ii) The actual passband frequency range where the response lies within ±10%,

  (iii) The actual passband ripple,

  (iv) The actual out-of-band rejection in dB,

  (v) The filter order,

  (vi) The number of multiply-adds per output sample using a cascade form realization.

(b) This part of the project concerns the computation of the average power for each channel. More specifically, we want to explore the possibility of sampling the output of each channel prior to computing the average power to reduce computation.

    To estimate the power in the ith channel signal of the OBSA, $x_i(n)$, at index n we will use the formula:

$$P_i^{avg}[n] = \frac{1}{N_p} \sum_{k=n-N_p+1}^{n} x_i^2(k).$$

The integer N_p is chosen to be large enough to extend over several periods of the quasiperiodic bandpass signal $x_i(k)$. Note that the precise value of N_p is not important, but it must be large enough so that $P_i^{avg}[n]$ tends to a constant value when the input power is constant. The equation for average power can be thought of as a cascade of a square-law device and an LPF as shown below:

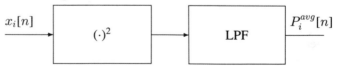

Obviously, the LPF is an FIR filter whose impulse response is:

$$h[n] = \begin{cases} \frac{1}{N_p}, & \text{for } 0 \leq n \leq N_p - 1 \\ 0, & \text{otherwise.} \end{cases}$$

The exact value for the cutoff frequency of the lowpass filter is not important since the role of this filter is simply to determine the DC value of $x_i^2[n]$. In fact, the finite bandwidth of the lowpass will be responsible for some errors in estimating the average power. One conclusion is that the larger the value of N_p chosen, the better the filter does its job.

    (i) Since the output of each bandpass filter is approximately bandlimited, show that the signal $x_i[n]$ can be downsampled by a factor of M_i *prior* to the calculation of the average power, without significantly changing the power estimate. (Note: There is a gain factor to be determined shortly.) Since the filters in the OBSA are not ideal, there is a source of error when downsampling and it will be necessary to make an approximation. State clearly the nature of the approximation that you have to make.

    (ii) Determine the largest M_i for each channel.

    (iii) How does the gain of the lowpass have to be modified to obtain the same value of average power as without downsampling?

(c) Since the output of each channel's bandpass filter can be downsampled, it might be advantageous to use FIR filters instead of the IIR filters determined in the previous section.

    (i) *Estimate* the lengths of the FIR bandpass filters to meet the same specifications used in part (a). *Do not actually design the filters.*

    (ii) From the estimated lengths, determine the number of multiplications and additions per output point required to implement the FIR filters. Take advantage of symmetry and the fact that the output will be downsampled. Do not include the operations used in computing the average power. Compare with your results in part (a) to see if there is any advantage in using FIR filters.

(d) In an attempt to ease the filter design constraints, especially for the narrowband filters, you can decimate (not downsample) the inputs to some of the bandpass filters. This can be done successively by factors of 2. The output of each decimator in the chain can then be filtered to provide the proper channel output. For the four channel OBSA, a cascade of three decimators should be adequate.

    (i) Show that, if the output of each "factor-of-2" decimator is used, the bandpass filters in each channel will have identical specifications. Thus, *one and only one* bandpass filter needs to be designed to realize all four channels.

    (ii) Determine the specifications of this bandpass filter and design the minimum length FIR filter to meet the specifications. Note that the specifications of this filter may have to be determined *in conjunction* with those of the decimation filter(s).

    (iii) The factor-of-2 decimators are all identical, so it is necessary to design only one lowpass filter for this decimation operation. Design this filter to be an FIR filter with minimum length. State clearly the specifications used in this filter design. Consider the fact that the filter response must not only perform an antialiasing operation, but also must satisfy the specifications on in-band ripple and out-of-band rejection when cascaded with the channel bandpass filters.

(e) Now consider the multirate implementation of the OBSA:

(i) Draw a block diagram of the system you created in part (d). Show all the individual filters and downsampling operators. Consider the possibility that the outputs of the bandpass filters can still be downsampled according to the strategy developed in part (b).

(ii) Determine the total number of multiplications and additions needed to implement this system. As before, take advantage of *every* opportunity to exploit symmetry and downsampling operators. Again, do not include operations used in computing the average power.

(f) It may be possible to simplify the realization a little more by imposing some constraints on the impulse response of the lowpass inside the decimator. Note that the ideal lowpass filter with zero phase and cutoff frequency at $\omega = \frac{\pi}{2}$ has the property that

$$h[2n] = \begin{cases} 0, & \text{for } n \neq 0 \\ 1/2, & \text{for n=0.} \end{cases}$$

(i) Show that the Fourier transform of any FIR filter satisfying the impulse response constraint above will have a symmetry around the frequency $\omega = \frac{\pi}{2}$. Describe this symmetry.

(ii) Using this symmetry, show how to specify the filter to your filter design programs so that you can guarantee that this condition will be satisfied. Verify that this strategy will work for both even and odd length filters.

(iii) Now make sure that you have specifications that will satisfy the requirements of the OBSA. Design the minimum length FIR filter for the decimator under the impulse response constraint above. Considering the length of this filter versus the length of the one designed in part (d), is there any computational efficiency to be gained by using this trick?

6.8 SINGLE SIDEBAND MODULATION

This project is concerned with the efficient implementation of a system for implementing single sideband (SSB) modulation. Such systems are used when several telephone channels are combined into a single high frequency channel using frequency division multiplexing (FDM) for long distance telephony. Assume that all of the signal information of interest in a telephone channel is confined to the frequency band between 300 Hz and 3300 Hz as shown in Figure 6.9. Outside this frequency band the signal spectrum is unknown. (You should not assume that it is zero.) Twelve different telephone channels are to be assigned to adjacent 4 kHz slots occupying the band between 60 and 108 kHz as shown in Figure 6.10.

In this problem, you should focus your attention on a single one of these channels—the one at a carrier frequency of 100 kHz. To preserve the signal quality for the signal of interest, the total distortion for frequencies in the 100–104 kHz range should be less than 0.5 dB. To limit crosstalk between channels, the out-of-band attenuation should be greater than 50 dB.

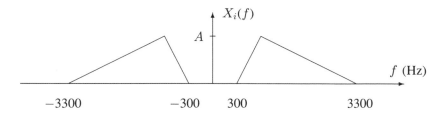

Figure 6.9. Spectrum of a telephone channel.

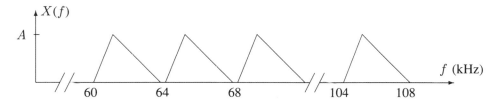

Figure 6.10. The FDM spectrum.

(a) The simplest approach conceptually is to simply multiply $x_i(t)$ by a 100 kHz sinusoid and pass the resulting signal through a highpass or bandpass filter. This is the approach suggested in Figure 6.11. For consistency with the other channels, assume that each channel is sampled at a sampling rate of 240,000 samples/sec.

   (i) Sketch a discrete implementation of this system.

   (ii) Determine the frequency of the discrete cosine and a set of design specifications for the digital bandpass filter that will meet the overall system specifications. Determine the orders of both IIR and FIR filters that will meet these specifications. Compare the number of multiplications, number of additions, and the memory required for their implementation and select the more efficient. *It is not necessary to actually design the filters.*

(b) The scheme illustrated in Figure 6.12 implements an SSB modulator by phase shifting. The filter $H(f)$ or $H(\Omega)$ (in radian frequency) in that figure is a Hilbert

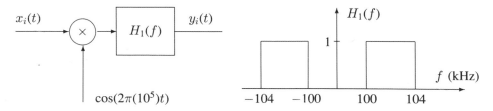

Figure 6.11. The modulation/filtering SSB modulator.

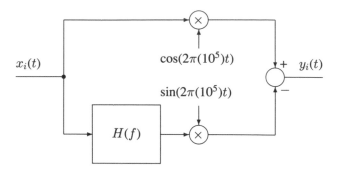

Figure 6.12. SSB modulation by phase shifting.

transformer, which has the frequency response

$$H(\Omega) = \begin{cases} -j, & \Omega > 0 \\ j, & \Omega < 0, \end{cases}$$

where $\Omega = 2\pi f$ rad/s. Design an SSB modulator that will meet the design specifications using this approach. The Hilbert transformer should be designed as an FIR filter. You are free to choose the design method to use, but as usual you should seek to minimize the total system complexity. To satisfy the requirements on out-of-band rejection you may need to add an additional bandpass or lowpass filter. If a lowpass filter is used before the modulation, you may choose to operate it at a lower clock rate and then to interpolate the result. You are also free to select the clock rate of the Hilbert transformer.

(c) A third system for implementing an SSB modulator is to build a digital realization of a Weaver modulator in which the modulation is done in stages. Figure 6.13 illustrates a Weaver modulator. The filters labeled $H(f)$ are lowpass filters with a cutoff frequency of 2 kHz.

   (i) Verify the operation of the Weaver modulator analytically.

   (ii) Determine the highest frequency of the outputs of the two signal multipliers on the left. Since this is considerably less than 200 kHz, the lowpass filters can be run at much lower sample rates than the final multipliers if signal interpolators are inserted at their outputs.

   (iii) Specify and design all of the components of the system. It is *extremely* important that the same filter be used in both branches of the modulator since the final summer produces a cancellation of unwanted parts of the spectrum. Select the clock rate for the lowpass filters to minimize the number of multiplications and additions in the overall implementation. Consider both IIR and FIR implementations of the filters and remember to protect against out-of-band energy.

(d) Which system, of all the variations that have been considered, seems to be the most efficient?

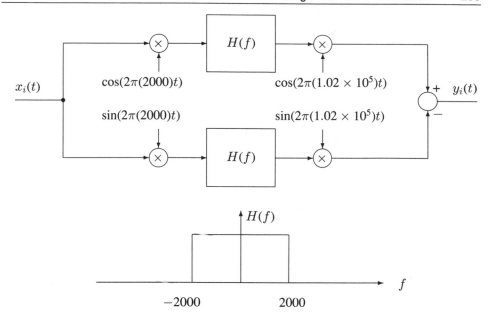

Figure 6.13.   A Weaver modulator for SSB modulation.

REFERENCES

[1] J. F. Kaiser, "Some Properties of a Family of Generalized Time-Limited Window Functions," *Proc. 1984 IEEE Int. Conf. Acoustics, Speech, Signal Processing*, paper 11.2.

[2] A. H. Nuttall, "A Two-Parameter Class of Bessel Weightings for Spectral Analysis or Array Processing—the Ideal Weighting-Window Pairs," *IEEE Trans. Acoustics, Speech, Signal Processing,* Vol. 31 (Oct. 1983), pp. 1309–1311.

[3] D. E. Dudgeon and R. M. Mersereau, *Multidimensional Digital Signal Processing,* Prentice-Hall: Englewood Cliffs, NJ, 1984.

[4] P. P. Vaidyanathan and T. Q. Nguyen, "Eigenfilters: A New Approach to Least-Squares FIR Filter Design and Applications Including Nyquist Filters," *IEEE Trans. Circ. Syst.,* Vol. 34 (Jan. 1987), pp. 11–23.

[5] W. H. Press, B. Flannery, S. Teukolsky, and W. Vetterling, *Numerical Recipes,* Cambridge University Press: Cambridge, UK, 1986.

[6] J. Candy and G. Temes, "Oversampling Methods for A/D and D/A Conversion," in *Oversampling Delta-Sigma Data Converters*, Candy and Temes, eds., IEEE Press: New York, NY, pp. 1–25.

[7] L. B. Jackson, *Digital Filters and Signal Processing*, Boston: Kluwer (1986), pp. 237–243.

[8] A. V. Oppenheim and R. W. Schafer, *Discrete-Time Signal Processing*, Englewood Cliffs, NJ: Prentice-Hall (1989), pp. 101–112.

Appendix: Software Documentation

The software provided on the disk in this book is a library of signal processing and filter design functions that provide the capability to design a wide variety of filters, study their implementation, and study their applications for processing signals. The first part of this appendix provides a brief description of each of the library functions. The second part describes the file format used to represent signals and filters. It provides the necessary level of detail so that you will be able to write programs in a language of your choosing that can interface with the DSP software.

A.1 QUICK REFERENCE FOR DSP FUNCTIONS

The most complete documentation for the DSP functions is accessible by the on-line **helpf** function. To use this feature, just type **helpf** for a list of the available functions. For a detailed description on the usage of each function, e.g., the function **function_name**, type:

helpf   function_name

This appendix provides an extremely abbreviated description of each function. By examining this section, you can gain a feeling for the functions that are available for your use. Each DSP function can be invoked by simply typing **f** followed by the function name. Similarly the graphics and display functions can be invoked by typing **g** followed by the function name. The function will then prompt you for all input parameters that it required from that point forward. In the overwhelming majority of the cases, it will be obvious how to respond after typing the function name.

f abessel Designs analog Bessel filters. The function arguments are the output filename and the filter order.

f abutter Designs analog Butterworth filters with a cutoff frequency of 1 rad/s. The function arguments are the output filename and the filter order.

f acheby1 Designs analog Chebyshev I filters with a passband cutoff frequency of 1 rad/s. The function arguments are the output filename, the filter order, and the ripple parameter ϵ.

f acheby2 Designs analog Chebyshev2 filters with a stopband cutoff frequency of 1 rad/s. The function arguments are the output filename, the filter order, and the ripple parameter ϵ.

f adaptfir Performs adaptive FIR filtering. The function arguments are the input file, the reference file, the initial filter coefficient file, the final filter coefficient file, the filter output file, the output error, and the step size μ.

f aelliptic Designs analog elliptic filters with a passband cutoff frequency of 1 rad/s. The function arguments are the output filename, the stopband cutoff frequency Ω_s, the passband ripple Δ_p, the stopband ripple Δ_s, and the filter order N.

f atransform Transforms an analog input filter $H(s) = B(s)/A(s)$ using the transformation

$$s \rightarrow \frac{c_0 + c_1 s + c_2 s^2}{d_0 + d_1 s + d_2 s^2}.$$

The function arguments are the input and output files and the transform coefficients c_0, c_1, c_2, d_0, d_1, and d_2.

f bartlett Designs a Bartlett window of length L. The function arguments are the output filename and the window length.

f bilinear Performs the bilinear transformation to convert an analog filter into a digital one or vice versa. The user is prompted for the bilinear transform variable β. The function can also perform the more general transformations

$$s \rightarrow \frac{c_0 + c_1 z^{-1} + c_2 z^{-2}}{d_0 + d_1 z^{-1} + d_2 z^{-2}} \quad z \rightarrow \frac{c_0 + c_1 s + c_2 s^2}{d_0 + d_1 s + d_2 s^2}.$$

f blackman Designs a Blackman window of length L. The function arguments are the output filename and the window length.

f cartesian Converts a sequence from polar to cartesian form, i.e., from a magnitude and phase representation to a real and imaginary part representation. The function arguments are the input and output filenames.

f cas Forms the cascade combination of two IIR filters. The function arguments are the two input files and the output file.

f cexp Multiplies a sequence by the complex exponential, $e^{\omega_0 n}$. If $x[n]$ is the input, then the output is $x[n]e^{j\omega_0 n}$. The function arguments are the input file, output file, and ω_0.

f convert Converts an IIR filter into a numerator file and a denominator file. The function arguments are the input filter, the name of the numerator output file, and the name of the denominator output file.

f convolve   Convolves two sequences. The function arguments are the two input files and the output file.

f diff   Designs a rectangularly windowed FIR differentiator prototype of length N with linear phase. The function arguments are the output file and filter length.

f direct1   Implements a direct form I difference equation with some explicit quantization and overflow features. The function arguments are:

> **input** is the input file;
> **filter** is the FIR or IIR filter;
> **out1** is the output file computed with floating-point arithmetic;
> **out2** is the output file computed with fixed-point arithmetic;
> X_{min} is the minimum quantization range value;
> X_{max} is the maximum quantization range value;
> **levels** is the number of quantization levels;
> **SW** is either 0 for saturation arithmetic or 1 for wraparound arithmetic;
> **output-length** is the number of samples in **out1** and **out2** to be computed.

f direct2   The same as **f direct1** for direct form II difference equations.

f divide   Divides the first input sequence by the second input sequence. The function arguments are the files for input1 and input2, and the output file.

f dnsample   Subsamples a sequence by a factor of M. The function arguments are the input file, output file, and M.

f dtransform   Transforms a digital input filter $H(z) = B(z)/A(z)$ using the transformation

$$z^{-1} \rightarrow \frac{c_0 + c_1 z^{-1} + c_2 z^{-2}}{d_0 + d_1 z^{-1} + d_2 z^{-2}}.$$

The function arguments are the input and output files and the transform coefficients c_0, c_1, c_2, d_0, d_1, and d_2.

f eformulas   A menu-driven function that evaluates three FIR filter parameter estimation formulas. The first is for the Kaiser window filter design and the second and third are for the Parks–McClellan algorithm. These formulas allow you to specify any two of three design parameters (ripple, transition width, filter order) in various forms. The function estimates the value of the remaining one.

f extract   Extracts a sub-block from a sequence. The function arguments are the input file, the output file, the input starting point, the length of the sub-block, and the output starting point.

f fft   Takes the FFT of the input. The input length must be a power of two and must begin at $n = 0$. The function arguments are the input file and the output file.

f filter   Convolves a sequence with an IIR or FIR filter and evaluates N samples of the output. The function arguments are the input file, the filter file, the output file, and N.

f gain   Scales a sequence or IIR filter by a constant gain factor G. The function arguments are the input file, output file, and G.

f hamming   Designs a Hamming window of length L. The function arguments are the output filename and the window length.

f hanning   Designs a hanning window of length L. The function arguments are the output filename and the window length.

f hilbert   Designs a rectangularly windowed ideal Hilbert transformer prototype of length L with linear phase. The function arguments are the output filename and L.

f histogram   Computes the histogram of an input sequence. See the on-line documentation for details on its usage.

f ideallp   Designs a rectangularly windowed linear-phase lowpass filter prototype with length L and cutoff frequency ω_0. The function arguments are the output file L and ω_0 (in radians).

f ifft   Computes the inverse FFT. The input length must be a power of two and must begin at $n = 0$. The function arguments are the input file and the output file.

f imagpart   Takes the imaginary part of the input sequence. The function arguments are the input and output files.

f impinv   Performs analog-to-digital filter conversion using the method of impulse invariance. The function arguments are the input file, the output file, and the sampling period T.

f kaiser   Designs a Kaiser window of length L with Kaiser design parameter α. The function arguments are the output filename, window length, and the Kaiser parameter α.

f kalpha   Computes the α parameter for the Kaiser window based on a set of desired filter design specifications. The function is completely menu driven and will allow you to enter the specifications in terms of the desired passband/stopband ripple, attenuation, transition width, and/or filter length.

f lccde   Implements a linear constant coefficient difference equation. Although the input and output may be specified on the command line, it is intended to be primarily menu driven. See the on-line help for more detailed information.

f log   Computes the natural logarithm of the input sequence. The function arguments are the input file and the output file.

f lshift   Shifts the input sequence by L, e.g., $x[n]$ is shifted to $x[n - L]$. The function arguments are the input file, output file, and L.

f mag   Computes the magnitude of the input sequence. The function arguments are the input and output files.

f matchedz   Converts an analog filter prototype into a digital filter using the matched z-transform. The function arguments are the input file, output file, and sampling period T.

f maxflat Designs maximally flat linear-phase lowpass FIR filters of the form

$$H(e^{j\omega}) = \frac{\int\limits_{-1}^{\cos\omega} (1+t)^p(1-t)^q dt}{\int\limits_{-1}^{1} (1+t)^p(1-t)^q dt}.$$

The function arguments are the output file, p, and q.

f median Performs N-point median filtering. N must be an odd integer between 1 and 99. The function arguments are the input file, output file, and N.

f multiply Multiplies two input files together. The function arguments are the two input files and the output file.

f nlinear Performs any one of three non-linear operations:
   1) $y[n] = \sin(\alpha x[n] + \phi)$,   2) $y[n] = e^{x[n]}$,   3) $y[n] = x[n]^\alpha$.
The function also has command line input capability. See the on-line help for further information.

f obutter Estimates the order of a normalized analog Butterworth filter given the stopband cutoff frequency Ω_s and the maximum passband deviation, Δ_s. The function arguments are the input file, the output file Ω_s, and Δ_s.

f ocheby1 Estimates the order and ripple parameter, ϵ, of a normalized analog Chebyshev 1 filter given the maximum passband deviation Δ_p, the maximum stopband deviation Δ_s, and the stopband cutoff frequency Ω_s. The passband cutoff frequency Ω_p is assumed to be equal to 1.

f ocheby2 Virtually identical to **f ocheby1** but for a normalized analog Chebyshev 2 filter. In this case, however, Ω_s is assumed to be equal to 1 and Ω_p varies between 0 and 1.

f oelliptic Determines the order of a normalized analog elliptic filter. The function arguments are the stopband cutoff frequency Ω_s, the passband ripple Δ_p, and the stopband ripple Δ_s.

f par Performs the parallel form implementation of two IIR filters. In other words, it adds two IIR filters together and puts them in direct form. The function arguments are the two input filter files and the output filter file.

f phase Computes the phase of an input sequence $x[n]$, i.e., $\arg(x[n])$. The function arguments are the input file and the output file.

f pksmcc A menu-driven Parks–McClellan filter design program. It designs optimal linear-phase FIR lowpass, highpass, and multiband filters. It also designs differentiators and Hilbert transformers. See the on-line help for more detailed information.

f qcyclic Performs wraparound quantization. It quantizes the input to the X_{min} to X_{max} range with L quantization levels. The function arguments are the input file, the output file, X_{min}, X_{max}, and L.

f quantize Performs conventional quantization. It quantizes the input to the X_{min} to X_{max} range with L quantization levels. The function arguments are the input file, output file, X_{min}, X_{max}, and L.

f rank   Performs rank order filtering. The function arguments are the input, output, N, and R, where N is the window length and R is the rank. The samples in the window are ordered from lowest to highest. Thus for $R = 1$, the minimum value is selected. For $R = N$ the maximum value is selected.

f realpart   Computes the real part of an input signal. The function arguments are input file and output file.

f reverse   Performs time reversal of a signal. For example, if $x[n]$ is the input, then $x[-n]$ is the output. The function arguments are the input and output files.

f revert   Merges two sequences $B(z)$ (input1) and $A(z)$ (input2) into an IIR filter $B(z)/A(z)$. The function arguments are the files containing input1, input2, and the output.

f rgen   Generates random number sequences of length N. The function arguments are the output file, N, and seed value. The seed is an integer used to initialize the random number generator. See the on-line help for more information.

f rootmult   Performs the inverse factorization operation. It converts a set of polynomial roots (i.e., the output of **f rooter**) into a polynomial. The function arguments are the input file and output file.

f rooter   Factors an FIR filter or sequence into its roots. The function arguments are the input file and the output file.

f siggen   A menu-driven program that allows you to design a variety of different signals such as square waves, sine waves, chirp signals, etc. In addition, it allows you to create your own files with user-specified coefficients.

f snr   Computes the signal-to-noise ratio for the signals contained in two input files $x[n]$ and $\hat{x}[n]$. The quantity calculated is

$$\text{SNR} = \frac{\sum_n (x[n] - \hat{x}[n])^2}{\sum_n x^2[n]}.$$

The function arguments are the two input files.

f subtract   Subtracts input2 from input1. The function arguments are files for input1 and input2 and the output file.

f summer   Computes the sum of the magnitudes and the sum of the squared magnitudes of the samples in the input sequence. The function argument is the input file.

f truncate   Truncates a sequence up to the index value L. The function arguments are the input file and output file.

f upsample   Upsamples a sequence by a factor of M. The function arguments are the input file, output file, and M.

f zeropad   Pads a sequence with zeros up to $n = L$, where L is a user-specified input. The function arguments are the input file, the output file, and L.

The graphics functions may be invoked by typing **g** followed by the function name. These functions may also be invoked using a command line call of the form

g   function-name   input-file

Some of these functions have a 2 in their name, such as **g view2** and **g slook2**. This means that they display two signal files simultaneously. The command line call for these functions is

g   function-name   input-file1   input-file2

Alternatively, these functions can be invoked by typing the function name and responding to the prompt for the input file(s). To exit these display functions, press the *"q"* key. This will return you to the operating system. Pressing the *"esc"* key will return you to the function menu. If the function does not display a menu, you will be returned to the operating system. The only exceptions are **g polezero** and **g apolezero**. These functions are completely menu driven and do not accept the *"q"* and *"esc"* key commands, command line input files, or command line parameters. The functions **g apolezero** and **g polezero** are invoked by simply typing their names.

g afilspec Displays the frequency response of an analog filter and allows you to examine the numerical values of Ω and the magnitude and log magnitude of $H(\Omega)$ using a cursor. The four directional arrow keys allow you to move the cursor across the screen with two different increments.

g afreqres Displays the frequency response of an analog filter. It provides options for many different displays including the magnitude, phase, real and imaginary parts, and group delay.

g apolezero A menu-driven program that allows you to read in an analog filter, simultaneously display the s-plane pole/zero plot and magnitude response, and to change the positions of the poles and zeros in the s-plane interactively.

g dfilspec Displays the frequency response of a digital filter and allows you to examine the numerical values of ω and the magnitude and log magnitude of $H(e^{j\omega})$ using a cursor. The four directional arrow keys allow you to move the cursor across the screen with two different frequency increments.

g dtft Displays the discrete-time Fourier transform of a digital signal or filter. It provides options for many different displays including the magnitude, phase, real and imaginary parts, and group delay.

g look Displays a sequence as a continuous waveform.

g look2 Displays two sequences, one above the other, as continuous signals.

g polezero A menu-driven function that allows you to read in a digital filter, simultaneously display the z-plane pole/zero plot and magnitude response, and to change the positions of the poles and zeros in the z-plane interactively.

g slook2 Displays two sequences as continuous signals superimposed on top of each other.

g sview2 Displays two sequences superimposed on top of each other.

g view Displays a sequence.

g view2 Displays two sequences at a time, one above the other on separate axes.

A.2 DSP FILE STRUCTURE

Two distinct forms of representation are available in a DSP file: 1) one for finite length sequences; and 2) one for infinite length sequences whose z-transforms are rational (IIR filters). Finite length sequences are composed of strings of numbers and have the form

$$b_0, b_1, b_2, \cdots, b_P$$

while IIR filters are composed of a ratio of polynomials in the variable z and have the form

$$H(z) = z^{-S} \frac{b_0 + b_1 z^{-1} + b_2 z^{-2} + \ldots + b_P z^{-P}}{1 + a_1 z^{-1} + a_2 z^{-2} + \ldots + a_Q z^{-Q}}.$$

The finite length sequence can be thought of as a special case of the IIR filter, where the denominator polynomial is a constant of value one. The file header and format accommodate both of these representations in a simple way. The first five entries of the file, which constitute the file header, are

N length,
P numerator order,
Q denominator order,
T 1–real or 2–complex,
S starting point,

where the parameters N, P, Q, T, and S are integers that have the following meanings:

N is the total number of coefficient entries in the file (excluding the header).

P is the order of the numerator polynomial. This value is only used for representing infinite impulse response (IIR) filters. In all other cases $P = N - 1$.

Q is the order of the denominator polynomial. This value is also only used for representing IIR filters. In all other cases $Q = 0$.

T is the coefficient type. $T = 1$ indicates real coefficients and $T = 2$ indicates complex coefficients.

S is the starting point (or index value) of the first sample of the signal.

The file structure is straightforward and can be illustrated by the two generic cases shown in Table A.1.

The files shown on the left and right are IIR filter and finite length sequence files, respectively, and depict the generic IIR and FIR signals. Several points are noteworthy here. First, the labels "NUMERATOR" and "DENOMINATOR" appear in the files to separate coefficients corresponding to the numerator and denominator of a rational system function. In the case of the finite sequence (**??**), the label "SEQUENCE" is included

Table A.1. File structures for IIR filters and sequences.

N	length	N	length
P	numerator order	$N-1$	numerator order
Q	denominator order	0	denominator order
T	1-real or 2-complex	T	1-real or 2-complex
S	starting point	S	starting point
NUMERATOR		SEQUENCE	
b_0		b_0	
$\vdots$		b_1	
b_P		$\vdots$	
DENOMINATOR		b_{N-1}	
a_1			
$\vdots$			
a_Q			

beginning of the sequence values. Second, T is most often 1 indicating real coefficients. When $T = 2$, the coefficients $b_0, b_1, \cdots, a_1, \cdots, a_Q$ are complex. Thus, each coefficient is a pair of values, the first corresponding to the real part and second to the imaginary part. A space (not a comma) is used to separate the real and imaginary parts. Third, the leading coefficient of the denominator polynomial is always assumed to be unity as shown in (A.7). Since the leading coefficient is always unity, it is not included in the file. The first term of the denominator explicitly shown in the DSP file is a_1. Finally, a signal file can always be examined by using the DOS *type* command to display it on the screen. The readable format of the file, which includes the descriptive labels in the file header, allows the parameters and all the signal information to be easily identified. In cases where you are generating the file header with a text editor or are writing a program that creates files, it is useful to know that the descriptive labels are not case sensitive. Labels may be either upper case or lower case. In fact, they do not even have to be spelled correctly in order for the DSP software to recognize the files. What is important is that spaces be placed between numbers and descriptive labels and that the number of words on each line be the same as shown in the examples.

Index